LEARNING TECHNOLOGY IN TRANSITION

LEARNING TECHNOLOGY IN TRANSITION

FROM INDIVIDUAL ENTHUSIASM TO INSTITUTIONAL IMPLEMENTATION

JANE K. SEALE (EDITOR)

Routledge
Taylor & Francis Group

LONDON AND NEW YORK

First published 2003 by Swets & Zeitlinger Publishers

2 Park Square, Milton Park, Abingdon, Oxfordshire OX14 4RN
52 Vanderbilt Avenue, New York, NY 10017

Routledge is an imprint of the Taylor & Francis Group, an informa business

First issued in paperback 2020

Library of Congress Cataloging-in-Publication Data

A Catalogue record for the book is available from the Library of Congress

ISBN 978-90-265-1963-5 (hbk)
ISBN 978-0-367-60465-3 (pbk)

Contents

Preface

Ten years ago, in 1993, the Association for Learning Technology (ALT) was set up to promote good practice in the use of learning technologies in higher education in the United Kingdom and to facilitate collaboration between practitioners, researchers and policy makers. This book has been written to celebrate the 10th anniversary of ALT and to chart the impact -past and future- of learning technology on post compulsory education. As one of the handful of individual enthusiasts who founded ALT ten years ago I thought I would provide an overview of the general context in which ALT was founded and how this links to the particular focus of this book: individual enthusiasm and institutional implementation.

Although on a much smaller scale, the ALT of a decade ago was not all that different from the ALT of today. The initiatives envisaged by its founder members were what might be expected of any organisation of the kind: a journal, a newsletter, workshops, an annual conference and if possible some mechanisms for influencing policy. The real differences between then and now lie not in the range of ALT's activities (even if that range has expanded and the activities have moved with the times), nor in ALT's objectives (even if the targets are more inclusive than they originally were), but rather in the capabilities of the technology and, closely linked with that, the way in which it has been exploited by those trying to promote its use within post compulsory education.

Even more so than now, user-expectations in the 1990s were consistently ahead of what the technology could deliver. In 1996, the film *Independence Day* presented its audiences with an extraterrestrial invasion of Earth finally foiled by plugging an ordinary laptop into the central computer of the alien mother ship. That may seem ludicrous today (especially to those of us who at the time were struggling with trying to link two ordinary terrestrial computers together), but it would appear that in the popular imagination of the mid-1990s such an action was perfectly feasible: no contemporary review of the film I have been able to locate mentioned the utter implausibility of that part of the story line. However, just as the language laboratory was at first perceived as a cure-all for Britain's linguistic infirmity, then found in practice to be wanting, so educational computing, which initially looked as if it represented the supremely effective solution to the problem of growing numbers of students, was soon seen to be deficient in a number of crucial respects when actually implemented. In 1993, when ALT was launched, software crashed just as frequently as today, and for the most part ran only on specified machines which not everybody had (these were the days of the great Mac-PC divide when the two types of 'serious' personal computers refused to talk sensibly to each other in the same way as some of their over-defensive apologists). Much worse, educational software was typically not up to the job, either because it had been written by a computer geek with no clue as to pedagogy or by a teacher who had learned the basics of programming a couple of weeks before deciding to put the newly learned skill into practice in the world of learning. In fact, in 1993 the gap between the limitations of both the technology and its exponents, and the effectiveness of CAL software when used by real students in real institutions, was huge. So huge that, looking back, I am surprised that any institution actually thought it worthwhile giving students access to it. And I am not standing under an umbrella while it rains on others: my own premature efforts at programming, which I believed at the time had some educational promise, now almost make me wince.

Yet the signs of things to come were, to some of us, plainly there. And herein is the link between the capabilities of that early technology and the attitudes of its champions as an educational tool. The point is illustrated with a bit of ALT's history. The Association's origins lie in some events of the year before its official unveiling. Graham Chesters and I were joint editors of a journal , *Interactive Multimedia,* which was bought in 1992 by a rival publisher who wanted to constrain us as editors by aiming the journal at computer scientists, while we were insistent that it should remain inter-disciplinary. We could not come to an agreement with the new publisher, and accordingly decided that we should start a new journal. When we informed the new publisher of our intentions, we were threatened with legal action should we attempt to launch a competing title, whether or not that title was aimed more widely than the confines of Computer Science. We contacted Jonathan Darby for advice, who, as it turned out, had been thinking about creating an association focused on learning technology. A short meeting of the three of us resulted in the decision to create ALT and to give it a journal, which would effectively replace *Interactive Multimedia* without competing directly with it. A few weeks later, the future founder members of ALT met for a day, and the Association was born. The point of this story is that the publishing company which had bought *Interactive Multimedia,* a well-established firm with experience of journal production, simply could not be persuaded that computing would interest a readership beyond computer scientists sufficiently large to guarantee adequate sales. Their view was that multimedia, which at the time was dependent on machines primitive by present-day standards and which consequently demanded of programmers that they squeeze blood out of a stone, was hardly for the non-technical, an attitude which may have had some validity but which also unmistakably spilled over more generally into the use of computers by "non-computerates". The founders of ALT saw things differently. We were confident that, even given the deficiencies of the available hardware and software, there would come a time when what we now call ICTs would be at the very core of higher education (further education came within ALT's remit only later), and in a wide variety of disciplines, even those traditionally thought of as requiring little more than pen and paper. The later ubiquity of desktop computers, with their hypermedia capabilities soon to be exploited to the full on the Internet, were to prove us right. In the meantime, we were mostly just a bunch of amateur enthusiasts, but enthusiasts who, while being well aware of the limitations of the technology, wished to see it exploited warts and all within higher education, and who were in no doubt that it would one day be completely reliable and universally accepted.

Has that day arrived? Not quite yet, but the movement is now surely unstoppable. ALT has moved from promoting the enthusiasm of a few individuals, as it was wont to do in its early days, to supporting the widespread institutional implementation we witness today. In 1993, very few higher or further education establishments took learning technology seriously; today there can hardly be an establishment that does not. Institutions are still full of individual enthusiasts, some of whom remain the principal driving force behind the implementation and exploitation of new technologies, but (for whatever reasons), those who control policy within these institutions have accepted that they must, at least, present an image of technological ascendancy when it comes to learning and teaching. That in itself is a major breakthrough about which the devotees of a decade ago could only dream: in those days many of them spent much of their time banging their heads against a sponge! It may be that there was an inevitability about learning technology becoming the vital part of post compulsory education it now is, something which would eventually have come about with or without the early commitment of individual aficionados. But whatever the case may be, ALT can surely be proud of the part it has unquestionably played by its very existence in accelerating awareness, and through its activities directly contributing to the present undeniably high status of learning technology within post compulsory education. We have moved from the particular enthusiasm of a handful of individuals to a general acceptance of the use of learning technology. This book therefore rightly celebrates a decade of achievement.

Gabriel Jacobs

1

Enthusiastic Implementation: Setting the Scene for Evolution and Revolution

Jane Seale

In April 2003, The Association for Learning Technology (ALT) celebrated its tenth anniversary and this book has been produced in order to commemorate this landmark achievement. The writing and production of the book represents a collaboration between old and new members of ALT as well as members of ALT's sister organisations in the Netherlands (SURF) and Australia (ASCILITE). The aims of the book are to use the topic of " institutional implementation" to present a review of the impact of learning technology on post compulsory education over the past few years; highlight and discuss key changes and developments that are shaping present and future activities and consider the implications for individual enthusiasts who work in the field of learning technology. This chapter will outline the context in which individual enthusiasts have operated and institutional implementation has occurred and introduce the main themes of the book. Four key themes are highlighted and discussed:

- The individual enthusiast and their role in institutional implementation;
- The institutional enthusiast and their role in local and global e-learning initiatives;
- Finding the evidence to justify enthusiasm and underpin implementation;
- Reinventing the individual enthusiast.

The Context of Individual Enthusiasm and Institutional Implementation

For the purposes of this book the definition of learning technology outlined by Seale and Rius-Riu (2001) will be applied:

> ALT understands learning technology in a broad conceptual sense as the systematic application of a body of knowledge to the design, implementation and evaluation of learning resources. The body of knowledge -the fruit of research and practice- is based on principles of learning theory, instructional design and change management. Learning technology makes uses of a range of communication and information technologies to support learning and provide learning resources.

Over the past ten years, three key factors have influenced the impact of learning technologies on post compulsory education (further and higher education):
- Technological developments;
- Policy drivers;
- Funding initiatives.

Technological developments

A review of any learning technology journal, book, newsletter or conference will reveal common phrases and beliefs about the status of technology today. To many we are now in a "digital world" where technological applications such as the Web, email and Virtual Learning Environments (VLEs) are so ubiquitous that we don't talk about them any more, while new mobile and smart technologies are emerging and catching our attention. The "digital world" is an exciting world, which offers those in post compulsory education opportunities to explore new ways of teaching and learning. In the digital world:
- Technology changes and develops at speed;
- Technology acts as a change agent "pulling" academics along in the pursuit of improving their practices;
- Life with technology is "rosy" and the equation is simple: technology plus student equals learning.

Whilst the chapters in this book refer to or describe the potential of technologies, they also explore how the contexts in which these technologies are being implemented can impact on this potential. These individual, departmental, institutional, national, international, social, economic and political contexts constitute the real world in which digital technologies are introduced. In the real world:
- Technology changes too quickly that that it is hard to keep up to date and there is little time to learn the lessons from our experiences of using "old" technologies;
- It takes more than the mere presence of technology to change practices;
- Technology can lead or drive change in practices but it might be at the expense of other factors;
- Life with technology can be messy and the equation is complex.

Technology changes too quickly
When technology changes too quickly it can be hard to keep up to date and there is often little time to learn the lessons from the experiences of using "old" technologies. In a review of the drivers that have influenced the implementation of learning technologies within further education, Wilson points to difficulties that can occur when the technologies change rather rapidly (see Chapter 5):

> The immaturity and volatility of some learning technology mean that there is a lot of work involved in keeping up with available products. Much effort has been wasted through poor understanding of the technology and its application.

While in Chapter 11, Oliver argues that:

> Learning technology often seems an amnesiac field, reluctant to cite anything 'out of date'; it is only recently that there has been a move to review previous practice, setting current developments within an historical context [...] Partly, this tendency to forget can be explained in terms of the speed of technological development. Nonetheless, many lessons learnt when studying related innovations seem lost to current researchers and practitioners.

The need to learn lessons is echoed by Conole who argues that little is understood about the impact that learning technologies have on organisations and individuals (see Chapter 10).

Technology on it own does not change practice
Littlejohn and Peacock, in Chapter 6, note how in the early 90's:

> There was, in many, a false assumption that exposure to computers and CAL packages was sufficient to drive the development of new forms of teaching with technology.

While in Chapter 7, De Boer, Boezerooy & Fisser note that despite having "one of the world's fastest and most advanced networks", Dutch higher education institutions are in no hurry to change the focus of their activities:

> Furthermore, no real dramatic changes in mission, profile or market position are expected, especially not with respect to new target groups like international students and lifelong learners.

Technology at what expense?
A commonly cited perception is that technology can lead or drive change in practices that might be at the expense of other factors such as pedagogy or academic freedom. Many chapters in this book refer to the tension that can exist between technology and pedagogy. This tension has led to arguments that curriculum development has been led by technology or that the way learning technologies are being used or applied is not

underpinned by sound educational principles. For example Oliver, O'Donoghue & Harper in reviewing the Australian context report that in the early 1990's (see Chapter 8):

> Moves at that time in the on-campus circles of higher education within Australia to adopt new technologies to supplement on-campus teaching showed symptoms of what is called technology-led curriculum development. This model of adoption of new technologies into education is still practiced in systems worldwide [...] and many questions remain in terms of its effectiveness and utility.

One particular focus for the tension between technology and pedagogy at the moment is the Virtual Learning Environment (VLE). For example, Littlejohn & Peacock argue that technologies such as VLEs have been implemented despite their "deficiencies in being able to support a wide range variety of educational models". While Conole warns of the dangers of senior managers being "beguiled" by hype and:

> [...] buying a Virtual Learning Environment to support learning activities and then decreeing that all courses must use the system without considering whether or not this might be pedagogically appropriate..

Jacobs in Chapter 9, goes beyond arguments about pedagogy to explore beliefs that technology, through its commercial sponsors, might be introduced at the expense of academic freedom:

> As powerful private interests, attracted by the image of a seemingly lucrative online education market, seek to use university standing to validate courses, senior managers, above all in research universities, are often torn between, on the one hand, the lure of private-sector money with its strings attached and its utilitarian, market-driven targets, and on the other the desire to maintain the time-honoured sovereignty and independence of the Academe.

However, in Chapter 2, Calverley argues against the common perception that the introduction of technology commonly overrides learning needs and implies that this perception leads to perhaps unnecessary resistance. These differences in opinion suggest that the implementation of learning technologies into further and higher education institutions may not always be simple or "clean".

Life is messy
In Chapter 6, Littlejohn and Peacock recognise the complications inherent in change associated with technologies:

> [...] the infrastructure is largely is now largely in place but technology in teaching is not ubiquitous and the vision is far from being realised [...] This is because technological issues have in the main been easier to solve than the more complex, social, cultural and organisational issues involved in mainstreaming technology in learning and teaching.

And there are that there are losses as well as gains:

> Letting go of familiar, comfortable practice and adopting new ways of doing this is
> a painful process, which all too often is ignored. Many staff development pro-
> grammes provide support within the cognitive and psychomotor domains […]
> However, programmes also need to reflect upon and take account of academics'
> anxieties to enable development within the affective domain.

In Chapter 10, Conole notes:

> Politics is a very strong theme that runs across all learning technology research. This
> in part relates to the over hyping which occurs, leading to an over expectation of
> what is possible. It is also partly due to different local agendas and associated in-
> fighting as well as the major impact that technologies can have.

The complexity of implementing learning technologies leads Oliver to warn us against
adopting rational or technical models of change and Conole to argue for the need to
develop a greater understanding of this complexity in order to develop more effective
practices.

Policy drivers

Over the past ten years a whole raft of reports on teaching and learning in post compulsory education have been published. Many of these reports have had an impact on government policy and consequential funding initiatives. The Dearing Report (NICHE
1997) has perhaps had the biggest impact on teaching and learning in higher education
and a number of chapters discuss this impact as it relates to learning technology (Chapters 5, 6, 9 & 11). If the introduction of learning technologies is "messy" then it is perhaps not surprising that there appears to be some inherent tensions in our response to
the policies that have influenced the landscape in which learning technologists work.
For some the tension lies in the fact that we have not met the vision set out by the policies. For others the tensions lies in the fact that our vision stretches further than that set
out in the policies and reports.

In Chapter 9, Jacobs discusses how the Dearing Report challenged learning technologists and institutions to take advantage of technological developments and capture
an international market of students. He comments on how lack of collaboration between
institutions has meant that "with only four years to go before the end of the post-
Dearing decade, it has to be said that we are far from reaching the point at which we
can sit back and be satisfied". Whilst Wilson pinpoints the Higginson report (1996) as
being influential in driving change in further education he comments that current " debates and initiatives still echo the Higginson Report while technology and many learners have moved on". This leads him to suggest that it would be interesting to see some
policy initiatives that focused more on the learner than the technology and started to set
out ways that learners could design and define their own learning.

Conole highlights the tensions that can exist between the needs of policy makers
and senior managers and academics and support staff on the ground level. She suggests
that policy makers are much more likely to be interested in potential efficiency gains

and cost effectiveness associated with learning technologies and will want to see evidence-based practice with a comparison of the benefits of new technologies over existing teaching and learning methods. This may however have a "potentially distorting effect on research in this field. Oliver warns against uncritical acceptance of policy but also acknowledges that there are occasions when policy " should be rallied to, rather than challenged".

Funding initiatives

The last ten years has seen a dramatic rise in the number of local and national funding initiatives that have been set up to increase the use of learning technologies within post compulsory education. A number of chapters in this book report on these initiatives and discuss their impact and influence. Higher education and further education funding councils in the UK have funded a wide range of initiatives over the past ten years including:

- Teaching and Learning Through Technology Programme (TLTP): See Chapters 4 & 6;
- The Fund for the Development of Teaching and Learning (FDTL): See Chapter 4
- ScotCIT: See Chapters, 4,5 & 6;
- The National Learning Network (NLN): See Chapters 3 & 5;
- Further Education Resources for Learning (FERL): See Chapters 3 & 5;
- The Learning Technology Dissemination Initiative (LTDI): Chapter 5 & 10.

Semi-governmental funding initiatives in the Netherlands (See Chapter 7) have led to the development of The Surf Foundation and the Digital University. Oliver, O'Donoghue and Harper describe how the Australian government funded a number of organisations whose aims were to increase the level of ICT uptake into curriculum delivery. In the UK and Australia early funding initiatives tended to focus on developing resources but eventually moved to focusing on institutional implementation. In the UK, The TLTP Programme is a prime example of (Baume & Jenkins 1996; Sommerlad et al 1999).

A number of chapters in this book consider the positive impact that these funding initiatives have had. For example, Oliver argues that:

> [...] we should not blind ourselves to the ways in which funding projects does lead to change. It buys time in which discussion, debate and learning can take place.

Conole builds on Oliver's argument that national funding initiatives have facilitated significant learning, in particular learning through experimentation:

> The research work, learning technology projects and developments, which emerged as a result of these national initiatives in general lead to an increased interest in the role of technology across education, senior management engagement, and consequential change in strategy. This provided opportunities to experiment using developmental funding.

While the chapter authors recognise the value of the national funding initiatives, they also highlight some important problems. In Chapter 4, Dempster and Deepwell present a review of national funded projects. In addition to highlighting problems with the short-term nature of funding they suggest that there can be a mismatch between what a project aims to achieve and the resources that they have to try and achieve these aims:

> National projects are not always resourced, equipped or best positioned to handle the task of organisational change as they generally have insufficient linkages with strategic planning and institutional mechanisms.

There is agreement by the authors that national funding has led to change, albeit in some cases, limited change. But comments by some of the chapter authors may lead us to question the depth of that change or the price that may eventually have to be paid by those in receipt of national funds. For example, Jacobs suggests that senior managers will 'chase the money' even if they do not necessarily buy into the causes that the money is being given for:

> The attitudes of administrators towards learning technology have unquestionably changed over the last ten years, but the change has been slower than might have been hoped for, at least partially because it has come about under steady govern- mental pressure (itself the result of public pressure) which translates into funding, rather than through a desire, however sincere, to accomplish a worthy mission.

While Wilson warns that a major consequence of this period of high profile investments will be that the funding bodies will look for "harder evidence of a return on this invest- ment." This will challenge many further (and higher) education institutions to sustain and embed the advances made through the use of national funds.

The Individual Enthusiast and Their Role in Institutional Implementation

All the contributors to this book would probably admit to being an individual enthusiast of learning technology For example, Wilson recalls how he started using learning tech- nology with his students ten years ago:

> I did it because it was exciting and new for me and the learners. Internet Technol- ogy offered them access to a wealth of resources that my college would not have been able to provide in other ways. A course web site enabled learners to access my classroom materials when they wanted. A chat room provided them with peer sup- port and an area where they could talk on or off topic as they pleased. I saw tech- nology as a mechanism to liberate learners, open minds and empower.

Like Wilson, I was excited at the opportunities that technology offered me to provide different learning experiences for my students. For example, I got involved in a project based in my institution called SCHOLAR (Maeir et al 1997) and developed a Micro- cosm application designed to encourage occupational therapy students to explore the applications that computers can have in the treatment of patients. I also explored the use of email and discussion lists to support the teaching on a first year psychosocial module for occupational therapy and physiotherapy students.

As individual enthusiasts Joe and I might be viewed in a number of ways. Jacobs might call us "amateur enthusiasts" (see Preface). Boyle & Cook might refer to us as "individual craftsperson" (see Chapter 3) while Littlejohn and Peacock might apply the term "pioneers". Oliver, O'Donoghue and Harper might define us as " early technology adopters seeking learning gains within their teaching". A less positive label that has been applied to individual enthusiasts is that of "computer geek" and other related terms. This label reflects the observation that some individual enthusiasts have focused on gaining and using their technical skills with little regard for the pedagogy of what they were trying to do (see Preface).

As an amateur enthusiast my attempts at using learning technologies in my teaching were motivated by educational rather than technical fascinations. My use of Microcosm represented an attempt to explore how a structured learning environment based on realistic case studies could encourage students to explore a resource base and find answers to self-posed questions. My use of email and discussions lists represented an attempt to promote and encourage discussion and reflection skills. I reported on my initiatives and the lessons that others could learn (Seale 1999a; Seale 1999b), but the publication that seems to have the most impact and influence on others thinking is that in which I report my relative failures (Seale and Cann 2000). In a sense I feel a bit like Jacobs, who in the Preface to this book confessed:

> my own premature efforts at programming, which I believed at the time had some educational promise, now almost make me wince.

The main reason why I wince at my attempts at using learning technologies with my students was that in hindsight I can see that it was not integrated enough into the curriculum, not just of my course but of the whole programme that the students were studying. I was also one of just two academics using learning technology within my department, and as McNaught and Kennedy (2000) report, this can be quite a lonely experience.

Despite my reservations about my own practices, I would argue that individual enthusiasts should be celebrated. Without their pioneering experimentation many current tools and practices would not exist. However, as some chapter authors in this book identify, the practices of some individual enthusiasts may place potential barriers in the way of wide-scale implementation:

- Their practices can be isolated which means many enthusiasts are "re-inventing the wheel", or replicating what others elsewhere have done (Chapter 3);
- Their practices are not always integrated into the mainstream practices of their department or institution (Chapter 4);
- They can capitalise on the practical skills they have gained to monopolise resources (Chapter 5).

The early chapters of this book chart the evolution of the individual enthusiast from the king of content to member of a community. In charting this evolution the chapter authors highlight how individual enthusiasts working together can help to embed learning technologies into their institutional practices.

The individual enthusiast as king of content

The original domain of the individual enthusiast was that of content designer. According to Littlejohn & Peacock these individual enthusiasts were:

> Early adopters who were skilled, knowledgeable, interested in technology and extremely self-reliant. These pioneers tended to be academics and this led to their use of technology being heavily contextualised within their individual subject disciplines.

This domain has however dwindled away as the role of content has changed. In Chapter 2, Calverley charts what she describes as the evolution of the form and purpose of content and discusses the implications this has for the roles of individuals such as tutors and academics. One of the recognised weaknesses of individual enthusiasts creating their own content for use with their students, is that in working in isolation, there is a risk that each individual will "reinvent the wheel", thus wasting effort and resources. This has lead to an emerging interest in whether resources created by one individual enthusiast can be re-used by another. In Chapter 3, Boyle & Cook discuss the design of learning objects that can be re-usable and pedagogically effective. They also examine the support that staff who are reusing learning objects will need.

The individual enthusiast as project innovator or champion

In additions to concerns over re-inventing the wheel, there were concerns that the practices of individual enthusiasts were isolated and not integrated into mainstream practices of their department or institution. Dempster & Deepwell evaluate the success of nationally funded projects in higher education that aimed to promote institutional implementation and move practice beyond that of individual enthusiasts, internally focused on their own interests. The move to institutional implementation did not mark the death of the individual enthusiast- instead a new breed emerged. The new enthusiasts had a wider focus and were essential to the success of institutional projects. They were either staff who had become specialised learning technologists employed directly by the project or lecturers who had their perspective transformed through participation in the project:

> These individuals were labelled as innovators in learning technology, recruited onto relevant committees who went on to influence policy and thus embed the ethos of the project within the institution

The individual enthusiast as a multi-skilled entrepreneur
Dempster & Deepwell conclude from their review of national projects that the rapid evolution of national programmes and support for teaching and learning development coupled with the emergence of institutional strategies for learning technologies has driven expansion of a vast composition of specialised staff working within the field of learning technology. Whilst these staff may have specialised technical skills, they also have to use a range of non-technical skills in their work including: curriculum development, negotiation, advocacy, research, evaluation, dissemination, project and team management, resource planning and trouble-shooting. In Calverley's terms these individual enthusiasts might be viewed as "multi-skilled entrepreneurs".

Wilson's review of the changes and drivers for learning technology in further education charts the development of the learning technologist from someone who: "completed internet modules, word-processing, PowerPoint and perhaps some desk-top publishing units and went onto monopolise the machines available in the staff resource base" to someone who: " has basic ICT competencies but can also share best teaching and learning practices with others".

In both Dempster & Deepwell's and Wilson's account of the emergence of the learning technologist there is recognition that they possess both specific and generic skills as well as technical and non-technical skills. This varied skills mix does mean however that a wide range of staff are working in learning technology and as Littlejohn & Peacock highlight, the term learning technologist is very broad. This has caused some difficulty for staff developers who are trying to find a focus for their efforts when catering for " someone teaching I.T. skills in how to use a VLE or an educational researcher."

The individual enthusiast as a partner or member of a community
In discussing the ways in which the LTDI in Scotland has been involved with supporting effective institutional change Mogey (1997) highlighted the fact that successful institutional implementation required a partnership between the institution's senior managers and individual enthusiasts:

> Institutional change requires partnership between a management, which recognises and responds to the resourcing implication of moves towards technology, and academic staff who can apply innovation with pedagogical insight, so that together good practice can be sustained for the benefit of all members of the post compulsory education community.

The chapters in this book pick up on this theme of partnership and community and explore its implications in a number of different contexts. For example, in discussing the reuse of learning objects, Boyle & Cook argue for a move away from individualism to a more collective approach to implementation:

> We need to move away from the model of the individual craftsperson all reinventing the wheel towards team-based collaborative development and reuse of high quality resources. There is a need for institutions to facilitate this co-operation and

avoid institutional barriers to progress. The ultimate and considerable challenge is to create vibrant communities of practice […]

Littlejohn & Peacock review the past successes of learning technology staff development and draw out emerging issues and trends that will impact on the staff development provision. They view the progression of staff development as a series of eras, which began with the pioneering era and is currently in a partnership era. In the partnership era, the role of staff developers is to bridge communities within and between institutions, where the communities reflect the emergence of new specialists, learning support professionals, academics and senior mangers.

The Institutional Enthusiast and Their Role in Local and Global E-Learning Initiatives

Just as individual enthusiasts can be readily identified in the learning technology field, so can institutional enthusiasts. So for example Sheffield College is associated with its Learning to Teach Online (Lettol) course, Durham University is associated with WebCT, Luton University with Computer Assisted Assessment (CAA) and DeMontfort University with the "electronic campus". These and other higher education and further education institutions have changed their 'identities' through their use and implementation of learning technologies. Many staff within these institutions have published accounts of the models and strategies that have been used to integrate and embed learning technologies (see for example McCartan and Hare 1996; Baume and Jenkins 1996; and Littlejohn and Cameron 1999).

In most accounts of institutional change there is recognition that successful institutional implementation of learning technologies depends on key individual stakeholders. For example, in reporting on their experiences at Durham, McCartan et al (1995) noted that the increase in institutional commitment to learning technology can be facilitated by the creation of a set of "new faces" to support its use. These "new faces" are appointed with institution wide briefs to integrate the efficient and effective use of learning technologies into the curricula. Whilst in 1997, Watson talked of the "ripple effect" that can result from the activities of individual enthusiasts within an institution. In describing institutional implementation of CAA at Luton, Bull and Zakrzewski (1997) noted that in an institution where the emphasis is shifting towards a student-centred curriculum, library and computing services staff play an increasing role in facilitating student learning. In reporting on the lessons learnt from the development of the "electronic campus" at DeMontfort University Brown (1998) stated:

> In order for innovation to successfully permeate the entire organization there have to be champions at all levels, ensuring an unbroken chain of commitment to the vision.

If individual enthusiasts pursue learning technology for the learning gains that can occur within their own teaching, what has driven institutions to pursue wide scale implementation of learning technologies, and in particular to develop an online presence? The common perception is that market forces have lead to institutions increasing the extent to which they offer online delivery to their home-based students. Those same market forces might also tempt some institutions to develop what Brown (1998) called a "virtual alter ago" and expand their sphere of influence by tapping into the international market for online education. As Jacobs in Chapter 9 comments:

> [...] the main institutional force for a change of direction will not be pedagogical, but hard-boiled educational politics entrenched in matters of finance.

In 1998, Brown felt that there were strong reasons why traditional face-to-face universities should resist the pressure to develop online delivery. In 2003, very few universities have reinvented themselves as players in the global e-learning market. The later chapters of this book explore the extent to which post compulsory education institutions in the UK, Holland and Australia wish to radically change their direction and "face the world challenge". In exploring the potential for this revolution the chapter authors highlight issues that will influence the extent to which institutions will collaborate with one another in order to compete in the global market.

The institution as a partner in local e-learning initiatives

In Chapter 7, De Boer, Boezerooy, and Fisser offer a Dutch perspective on institutional implementation. They note that many institutions have moved beyond a pioneering stage to institutional-wide managed change in a bottom-up and top-down approach. Although institutional collaborations do exist in Holland, (e.g. The Dutch Digital University- a consortium of ten HE institutions) De Boer, Boezerooy, and Fisser argue that these collaborations are motivated by a desisre to strengthen the "home-base" for Dutch students rather than to develop a virtual base for international students.

This rather conservative approach to online delivery appears to be influenced by lecturers and academics who wish to preserve the nature of the relationship that they have with their students:

> […] the classroom orientation model is the most common model used within Dutch higher education; a model in which instructors and other actors highly value the face-to-face interaction and direct communication between instructors and students and among students.

The institution as a partner in global e-learning initiatives

In Chapter 8, Oliver, O'Donoghue and Harper trace the development of Information and Communication Technology (ICT) in higher education in Australia and note a strong record of distance education and flexible learning opportunities. Unlike Holland, a number of Australian universities have joined or helped to form a range of consortia or alliances aimed at seeking new markets and competing on a global scale (e.g. Universitas 21 or the Global University Alliance). Oliver, O'Donoghue and Harper suggest that as Australian universities change the way they deliver teaching and learning opportunities, their internal structures will also change from "old-fashioned command and control" systems to more informal flexible and dynamic systems.

In Chapter 9, Jacobs discusses the extent to which UK higher post compulsory education institutions have collaborated to tap into the world-wide e-learning market by producing joint e-learning initiatives. In stark contrast to the individual enthusiasm highlighted in earlier chapters, Jacobs paints a picture of individual resistance, particularly from senior managers who are apprehensive about what they perceive as a risky investment and paying lip service to the Internet "without a deep commitment to seeing it used". This resistance or apprehension is due in part to an uncertainty about whether the financial risk will be justified or balanced by a real market demand for online courses.

Finding the Evidence to Justify Enthusiasm and Underpin Implementation

In Chapter 10.Conole provides an overview of current learning technology research and outlines the range of current e-learning research questions and issues. She argues that e-learning research is important because we need to understand the impact of learning technologies at an institutional and individual level.

> [...] learning technologies now have a significant impact at all levels of universities and colleges, from organisational and structural issues, through changing the nature of roles and functions and of course the impact on learning and teaching. Little is understood about these processes and how they are changing. There is a need to research these in order to better understand the associated issues and draw out lessons learnt and examples of good practice to inform future developments.

Conole also talks of needing to encourage academics to become involved in learning technology research alongside their mainstream subject-based research. If the learning technology research is not valued and recognised as equal then it is unlikely that academics will be prepared to invest time in this or be prepared to divert attention away from their other activities.

The call for individuals to conduct research that might address the needs and questions of their institutions raises an interesting tension: the tension between faith and evidence or experimenting and experiments. Brown (1998) highlights this challenge by noting that individual experimenters need "faith in the probability of desirable outcomes" while institutions require "convincing evidence" if they are to sanction wholesale institutional adoption of technologies. The drive for evidence does not always allow time for the kind of "experimenting" that individual enthusiasts might carry out where the potential for failure is as great as the potential for success. In Chapter 11, Oliver argues that failure is a necessary part of the learning and theorising process. In other words, experimenting gives us the opportunity to explain why the application and implementation of learning technologies are successful and why they are not. Butler (1997) argues that institutions should embrace such experimentation and " treat the opportunity in the same way they treat the opportunity to establish a research centre".

Reinventing the Individual Enthusiast: A Salute to the Young Turks of the Future

Earlier in this chapter the different ways in which individual enthusiasts have been viewed over the past ten years were discussed. This ranged from "amateur enthusiast" to "computer geek". A review of the contributing chapters in this book leads me to suggest that the rise in wide-scale institutional implementation has not led to the demise of the individual enthusiast. The original enthusiast might have been focused on their own localised teaching contexts and been marginalised by the their institution, but today's enthusiast is firmly located within the heart of the institution as policy champions and project innovators (Dempster & Deepwell, Chapter 4).

But what about tomorrow's enthusiast? In Chapter 9, Jacobs appears to identify a new breed of academic:

> Academics are hardly thought of with warmth and generosity by Government officials, nor by the public at large. Especially as regards the so-called upper-echelon universities, they are seen by many as arrogant, resistant to change, living at the top of their ivory towers with their heads in the clouds. That view may be undergoing something of a transformation as young Turks with bright, so-called relevant ideas enter the profession.

Here, the term "Young Turk" is an interesting one. Jacobs does not use it in a necessarily complementary way. Indeed, a dictionary check reveals that "Young Turk" means "troublesome person". However, throughout this book there are hints that learning technologists should create a bit of trouble in the sense that they should not accept without question certain aspects that are affecting the world in which they work. For example, Wilson challenges the growth of 'pedagogy poor' applications of technology:

> I groan at the thought of students faced with death by PowerPoint both in the lecture theatre and now in the virtual learning environment. They will also really 'enjoy' the endless automatically generated emails they will be sent asking why they have failed to submit their assessments on time.

Conole urges researchers to become involved in learning technology research alongside their own subject-based research and find (amongst other things) the evidence that may dissuade senior managers from making unwise decisions about the use or value of technologies. Whilst Oliver urges learning technologists to take a stand against what he sees as the "rhetoric of managerialist discourses of efficiency and control".

Indeed, Oliver himself might be a good role model for the 'Young Turks' of tomorrow. In his review of the chapters within this book he does not always agree with what the other authors have written and has some strong personal views about the ways things should move in the future. However, he justifies his position, through reference to theory and literature and is reflexive in the sense that he fully acknowledges his personal biases and agenda. The learning technology community has developed to such an extent that it can probably handle differences in opinion, but whether it is operating in an environment in which it can tolerate a rise in the number of "Young Turks" who 'push the boundaries', embrace failures as well as successes, deliver control of the learning experience over to the students and forge alliances with "Young Turks" from competing institutions remains to be seen. This might be, then, the challenge for the next ten years.

Conclusions: Evolution or Revolution?

The chapters in this book paint a picture of an evolving landscape in the field of learning technology where evolution is understood in the context of growth and development. In Chapter 2, Calverley charts an evolution in the role of content, while in Chapter 3, Boyle & Cook call for an evolution in the role of the content provider. Dempster & Deepwell, in Chapter 4 describe the development of national funding initiatives in higher education and the impact they have had on both institutional implementation and the roles of individual enthusiasts. In Chapter 5, Wilson describes how further education has developed, through an extensive programme of funding and support initiatives from a position where accessing a PC may have been problematic to a position where most colleges have been able to develop virtual learning environments and "start to wrestle with the creation of an managed learning environment (MLE)". Littlejohn & Peacock in Chapter 6 chart the changing role of staff development over the past ten years. In reviewing Australian developments Oliver, O'Donoghue and Harper in Chapter 8 conclude:

> The picture that emerges in the Australian post compulsory scene of ICT in education is one demonstrating significant growth and change and continual evolution of ideas and applications

Whilst Conole, in Chapter 10, seeks to explain how the field of learning technology research has evolved in line with typical development stages.

That the learning technology landscape is evolving will probably not be a surprise to readers, but what may be a surprise is that the chapters cited here appear to be painting a picture of gradual change rather than radical change. The last ten years have not been particularly characterised as one of revolution and some chapter authors no not predict a revolution in the next ten years. For example, in Chapter 7, De Boer, Boezerooy & Fisser (2003) report that Dutch higher education institutions do not expect any revolutionary change as a result from or related to the use of ICT. Whilst in Chapter 9, Jacobs argues:

> It was John Kenneth Galbraith who said that there are two classes of forecaster: those who don't know and those who don't know they don't know. We should be wary of the latter, and accept only with a dollop of suspicion self-assured predictions that a global educational revolution based on learning technologies is just around the corner and will therefore automatically attract huge 'defence' spending.

So whilst some might persist in talking about the revolutionising of learning in general and e-learning in particular, a review of the past ten years of learning technology developments and what the next ten years might hold leads me to a more cautious prediction, involving what Oliver in Chapter 11 terms "dynamic evolution".

References

Baume, C., & Jenkins, A. (1996). IT term: a model for institutional change? *Active Learning*, 5, 53-55.

Brown, S. (1998). Reinventing the University, *ALT-J*, 6,3,30-37.

Bull, J., & Zakrzewski, S. (1997). Implementing learning technologies: a university-wide approach. *Active Learning*, 6,15-19.

Butler, J. (1997). What is more frustrating: achieving institutional change or herding cats? *Active Learning*, 6,38-40.

Higginson, G. (1996). *The use of technology to support learning in colleges. Report of the Learning and Technology Committee (Higginson Report)*. London: Further Education Funding Council.

Littlejohn, A., & Cameron, S. (1999). Supporting strategic cultural change: The Strathclyde Learning Technology Initiative as a model. *ALT-J*, 7,3,64-74.

Maeir, P., White, S., & Barnett, L. (1997). Using educational development strategies to implement learning technologies. *Active Learning*, 6,10-14.

McCartan, A., & Hare, C. (1996). Effecting institutional change: the impact of some strategic issues on the integrative use of IT in teaching and learning. *ALT-J*, 4,3, 21-28.

McCartan, A., Watson, B., Lewis, J., & Hodgson, M. (1995). Enabling learning through technology: some institutional imperatives. *ALT-J*, 3,1, 92-97.

National Committee of Inquiry into Higher Education (NCIHE) (1997). *Higher Education in the Learning Society*. [On-Line]. Available: http://www.leeds.ac.uk/educol/ncihe/

McNaught, C., & Kennedy, P. (2000). Staff development at RMIT: bottom up work serviced by top-down investment and policy. *ALT-J*, 8,1,4-18.

Mogey, N. (1997). LTDI: Supporting successful implementations of learning technology. *Active Learning*, 6,27-29.

Seale, J., & Rius-Riu, M. (2001.) *An introduction to learning technology in tertiary education in the UK*. Oxford: Association for Learning Technology.

Seale, J.K., & Cann, A.J. (2000). Reflection on-line or off-line: the role of learning technologies in encouraging students to reflect. *Computers & Education*, 34,309-320.

Seale, J. (1999a). Introducing new learning technologies into the Occupational Therapy Curriculum: Evaluation of a Hypermedia Package. *British Journal of Occupational Therapy*, 62,4,144-150.

Seale, J. (1999b) Learning technologies and the lifelong learner: armament or disarmament? *ALT-J*, 7,1,61-67.

Sommerlad, E., Pettigew, M., Ramsden, C., & Stern, E. (1999). *Synthesis of TLTP Annual Reports.1999*. Unpublished report, Tavistock Institute.

Watson, B. (1997). Supporting the integration of IT into the curriculum at the University of Durham. *Active Learning*, 6, 34-37.

2

Reconsidering the Role of Content: Technology, Learning and Learning Technology

Gayle Calverley

Bluntly put, in today's clichés, 'The jury is out on whether 'content is king''. The role of content within higher education courses is rapidly now being relegated to the position of a background resource that supports enhanced provision of learning opportunities. This chapter will explore the context in which this dramatic shift has occurred over the past ten years and the implications this has for institutions and the individuals that work within them.

The Age of Digital Technologies

The changes in the role of content have largely been driven by the coming of age of digital technologies, to the stage where they can realistically be employed within learning in a transferable fashion. Successfully managing the change in presentation of courses to alternative means requires effective analysis of the purpose and nature of the interventions offered. This re-analysis of core purpose, coupled with the functionality offered by new developments in technology, is significant for those involved in learning. The advantage of this combination lies in the greater flexibility allowed for design and presentation of content, so that it can be positioned for maximal impact in the learning process and congruent with other necessary aspects of learning. It can now fit naturally into the process of knowledge and skill acquisition, and the application of learning. In the ideal situation, it can also be freed from constraints imposed by those non-serviceable barriers of management and presentation previously imposed by linear media and processes. Improvements can continue to be achieved into the future through adoption of emerging technologies, where these are selected and applied as carefully thought-out adaptations to stimulate learning.

A Change in Pedagogical Perspectives

Although the employment of digital technologies has increased, much of the open commercial e-learning market is deemed to be of questionable value and quality by educators, particularly where the continued focus of these offerings is based on content presentation as the key or sole component for online learning. This arises from the viewpoint that learning is in fact based around activities that encourage the development of critical thinking skills (Kreber, 1998;Gadzella et al, 1996), reflective practice (Reynolds 1998; Page and Meerabeau 2000), personal development through changing perception (Page and Meerabeau 2000; Knight 2002) and induction into one or more communities of practice (Tiwana & Bush, 2001). Taken in this context, content presentation as the key or sole component for online learning does not generally, by definition, offer any opportunity to develop these capacities for learners who have no initial skill basis or relevant community background. At best, it assumes the learner can apply those skills they have already developed as a successful Lifelong Learner. At its worst, it thwarts even the best attempts of those skilled in this area.

As a consequence of this perspective shift, other aspects inherent in learning have, quite rightly, been drawn to the fore. As a learning community, this has allowed us to begin to provide the type of learning experience we would all like to participate in when it comes to our turn to work this way. We still have a long way to go, yet the fact that we might even consider working like this beyond our initial studies and career establishment is also indicative of change and (to quote from the coffee room) heralds 'The Age of the LifeLong Learner'.

Content as a Stimulus to Learning

Although perceptions of the role of content are changing presentation of content still offers an effective starting point for the process of learning. It helps the learner formulate questions and to begin discussion from a common position with a peer group and the tutor. This can be developed with a view to working towards meeting the defined learning objectives of a course. The process following the presentation of content demonstrates that, in many respects, active learning can parallel resource-based learning. For this parallel to be successfully achieved relies not only on the availability of good "fit-for purpose" resources, but also on their effective integration into the learning process (Calverley & Shephard, 2003). Clearly, the term 'content', used in this sense, can be an ambiguous and open term. For example, how far does it define the resources themselves, and how far the packaging that the tutor builds around these for effective presentation to his or her students? However, it is worth noting that the parallel described here is not largely affected by the interpretation chosen, rather it affects the number of variables to examine within any given learning process, and their individual attributions to the success or otherwise of the intervention.

Conversely, once the skills for Lifelong Learning are obtained, the reverse process is facilitated. That is, the study of diverse sources of content can assist the learner to develop new learning objectives of their own. In this case, these new objectives are likely to be related to specific personal development or current business needs, and occur in part from developing the kinds of changes in viewpoint that result from invoking reflective processes as part of everyday professional practice. In this context, a strong peer and/or mentor network can speed this process, acting in a similar fashion to the tutor in a formal course environment. This assumes cases where the learner's critical faculties are sufficiently advanced to be able to draw informed conclusions from the evidence and (or) information available. This is a conceptual extension of the views of Rader (1995) and Candy (1988), who examine information literacy as a critical skill for the Lifelong Learner, and of McDowell (2002) who suggests that encouraging information literacy in students shifts towards increased student independence in learning.

The Influence of Standards on Content

Even within well-designed courses and training material, context specificity can draw accepted views of learning into question, causing reconsideration of the value of non-interactive or non-discursive content. This becomes significant for areas of training where linear sequencing of content presentation and testing is vital to skill development; and where there is little room for deviation from a tried and tested path to be successful, and in many such cases, remain safe and alive. Yet its apparent value and relevance for other educators is limited due to the inappropriateness of transfer of such techniques to areas of learning that are inherently non-linear, or differently structured, by nature. The alignment of such approaches is familiar to those of the technological standards community with a special interest in learning, and in the development of systems to support learning so that different approaches and learning needs can be accommodated within such systems.

One example of this type of situation is the relationship between the IMS specification for 'Simple Sequencing' and the IMS specification for 'Learning Design'. 'Simple Sequencing' allows for basic forms of sequencing to be applied to materials within a training context, and for this to be transferable between systems. However, this approach to ordering content is not very useful, for example, where multiple purpose alternative paths for the learner are required. Instead, this type of approach is offered through the IMS 'Learning Design' specification. In fact, both can sit quite comfortably together within the full spectrum of learning practice, and 'Simple Sequencing' can be drawn on within 'Learning Design' applications as and when it is appropriate, under those conditions containing a specific subset of learning requirements that require this approach. Neither dictate the form in which a course is to be presented, they merely allow the educator-developer more effective ways to present the learning choices online, of the type that good learners already naturally invoke at point of need to reach their learning objectives. However, as specifications evolve, there is a constant search for understanding within the community [1] as to how various components of the stan-

dards, specifications, and reference models (e.g. IMS, SCORM, IEEE LOM), fit together to support the ways in which we can encourage learners to perform at their best
in these new types of environment.

Changes in the way Higher Education Works With Commercial Partners

The scenarios outlined in the previous sections have led, on the technical side, to development in the areas of educational repositories, reusable learning objects and interoperability between educational systems, processes and institutional policies. These developments have drawn together education and commercial partners in new ways related to the exploration of learning, educational processes, and content handling.

This change in the way education is working with commercial partners has an impact on the perception of content, especially in the way we approach delivery across
multiple systems and groups. For example, communicative and some assessment functions supporting online learning may be offered through the functionality of a Virtual
Learning Environment. However, it is still the responsibility of those involved in course
design and delivery to determine how the content is procured, developed and handled
during the creation of a successful course. This has led to multi-component learning
systems that are able to link related content, communication and administrative functions, including Virtual Learning Environments (VLE), Repositories, and Content
Management Systems (CMS). In addition, institutional management functions are often
aligned through an integrated Managed Learning Environment (MLE) structure.

Reducing development loading through custom building

Although the cost models of products developed with commercial partners are often
high, in effect, this distribution of producers and providers has resulted from the necessary and gradual devolution of development labour and production costs from the institutions themselves. This allows institutions to concentrate on support for their own staff
and courses rather than software development, and on those areas of system customisation that must be managed in-house. The commercialisation of educationally oriented
delivery systems for content, resource, and communication has been a critical development in allowing online learning to become manageable and scaleable for institutions.

It is interesting to note the diversity in the market, with key commercial players
such as IBM and Microsoft offering rollout solutions for education, alongside commercial vendors that have arisen from developments originating within an institution. Yet,
initially, the technologies being developed commercially by non-education based companies, in attempt to break into the educational market, were functionally unsuited to
the needs of the sector, and those designed for one part of the world did not necessarily
translate well to other continents, linguistically, functionally, or technologically.

A number of today's commercial developers are companies originally derived from
the international higher education sector (e.g. WebCT, Intrallect, and Extensis). Yet
once commercial and as the business builds, some of the drivers that make these products effective as a first development are lost. Even for institutionally driven companies,
once the commercial base is established, working and product priorities often become
sufficiently changed, making it difficult to retain an up-to-date hands-on understanding

of client difficulties. Although close client relationships are an accepted way around this, circumstances often remain problematic for institutions in terms of product customisation to specific needs. This means there is still a great deal of co-working and collaboration needed for these products to fully meet the needs of users, and to maintain pace with changing sector needs.

In terms of content, this manifests itself in the delivery of products that impose a particular or assumed approach to learning around which tutors must then work and in constraints for interoperability and transfer of materials between systems. A third issue, relating to pre-purchase commercial content, involves the level of re-engineering required to embed or convert such material to alternative contexts or cultural environments. This can be mediated depending on the origin and nature of the material (such as commercial partnerships for marketing of specifically tailored contextual material). Cultural differences in accepted practice can exacerbate the problem for products that cross continents.

Although business and educational needs still conflict so that it is difficult to achieve this seamlessly, there are now products that are able to adequately, if not perfectly, perform this role, and the communication lines between vendors and institutions are open significantly wider than before.

Interoperability Versus Reusability

Good software design practice recognises that software must be conceived of as something far broader than program code. It must include all design specifications, user documentation, test cases, data, codes and human interfaces (Gilb, 1998). Therefore, when considering system support requirements across the entire function of an educational institution, or in other words at the enterprise level, issues of interoperability and the reuse of learning materials and course content become highly significant. In this context, architecture refers to the high-level answer to the many quality and resource requirements of a system so that it is possible to achieve reliability, and to meet those quality and resource requirements across the integration of all its functions. Due to the full and integrated coverage implicit in an 'Enterprise Architecture', it is recognised that its development must in turn be based on the institution's strategic vision (Malhotra, 1996). Consequently, content handling, delivery, distribution, and management need to be considered in terms of how these processes naturally follow from their integral links to learning, rather than the other way around. If this does not happen, and if content processes are not taken into account within design requirements and goals, then one of, or arguably, *the* most significant aspect(s) on which institutional business is based (i.e. learning), becomes omitted.

This is clearly important in an environment where the ability to share materials and resources across an institution's systems, for delivery across a range of courses as well as a range of student locations and circumstances, is key to effective delivery of online and distributed learning. Examples include a member of staff who wishes to reuse sections of his or her own material with several different student groups, yet students groups common to that member of teaching staff may not all use the same Virtual Learning Environment (VLE) depending on where their course is accessed. Alternatively, an institution may change its preferred method of delivery for online courses in a particular area, and may seek techniques for their staff to be able to transfer course material between systems in a straightforward manner. Further, there may be staff working within a team-teaching environment who need to be able to share the course material and resources between themselves and the students.

While in many senses, interoperability and reusability are linked in educational and learning contexts (typified by the reusable learning objects debate outlined in Chapter 3), material designed to be interoperable may not be reusable, and vice versa. Interoperability derives from the ability to share information, resources, and other material between different systems and to enable such material to be easily transferred between technologies, either in the context of upgrades, or in situations of technology transfer and/or knowledge sharing. Interoperability for educational systems has largely been driven by standards derived for learning situations and provides mechanisms by which content can be shared across systems or presented to users. Reuseable material on the other hand, holds much of its value in its ability to transfer between learning contexts as opposed to between delivery systems. As such, it is the contributor or content author who controls the form of the content and its potential for reusability across different learning scenarios. In addition, reuseable material may be non-technology based. In the examples above, it is assumed that the member of staff already has workable scenarios for reuse based on their own materials, and so the 'Enterprise Architecture' is more concerned with interoperability. However, implementation of an image system for use across disciplines, as part of that architecture, for example, is likely to force an additional range of educational reuse conditions onto the design of the architecture. In either case, design of materials for reuse and the organisational implications for change relating to architecture design will feature to some extent.

New Learning Opportunities From Changes in Content Delivery

Technology is recognised as one of the single largest change agents, alongside major sector changes in structure and resource modelling, which has forced a rethink of institutional core business and practices over the last ten to fifteen years. This makes it worth continuing to consider the options becoming open to us through emerging technologies and how they may help us to realise some of our other 'desired possibilities'. This will allow us to offer more effective associations of learning with skill and practical based development, and its application to the real world environment. This is timely, in a period where consolidation of what is learned in training is greatly reduced for those real-world situations where decisions need to be made much more quickly than in the past, and frequently with incomplete or little accurate information available (McHardy &Allan 2000; McHardy and Henderson, 1994). Yet there is still a common perception that introduction of technology is frequently made in such a way as to override learning needs, which provides resistance to this exploration in certain environments.

Developments in the area of mobile technologies are already offering significant possibilities for offering content and in-context information to provide more greatly tailored learning circumstances and point-of-need feedback. The benefits of instant access have already been demonstrated through the use of flexible workspaces created using wireless networks. For example, engineering students at the University of Twente (de Ruijter & Maas, 2003) have commented on the ability to directly compare information at point-of-need during project work. This minimises error, as exact original information is readily available so that no assumptions are required. Key information can be retrieved readily from its source regardless of location (within the range of the available networks). This maximises potential for learning as there is no 'going back to check', and students can move directly on with the next stage of work. This reduces iterations required to complete a given task, maintains flow of thought, and also allows more effective student-tutor interaction based on more accurate information. In this way, content, and its delivery, becomes so naturally integrated into the main learning activity that in some senses its presence is ephemeral against task context. In part, this is also due to the way content information is reinforced during task-oriented work, so acting as partial feedback. Students will have made an attempt to work from first principles and recall before referring back for correctness of their response.

Other areas where instant access models are being exploited include developments to support learning conversations using personal technologies. Work by Sharples et al (2002) effectively exploits the combination of mobile technologies with concept maps to allow learners to guide their conversations, with a more experienced peer, tutor, or colleague, from the edge of their existing knowledge towards the topic areas that are most useful for their current needs. To do this, the learner creates a concept map on a mobile computing device while conversing with a (relative) expert. The device also holds a complete map of the topic, which is initially hidden from view. As the conversation progresses, the learner creates nodes and the device compares these with the hidden map and displays matching nodes with their immediate links. From this, the learner can then extend the discussion in an appropriate direction. The focus at present is for learning outside of formal teaching, for example where one adult learns from another as they discuss a topic together, especially over the telephone, and take notes.

Further interesting in-context examples exist which presently make use of mobile situations for casual learning and information orientation. Location Activated Nomadic Discovery (LAND) offers location-sensitive visitor and navigation information for visitors to Cumbria, using portable devices and mapping technologies (Taylor et al, 2002). By relating user profile to current location to content, visitors are able to discover 'hidden' environmental and cultural knowledge related to the region. Developments in wearable computers, such as the χ^3 and WECA PC (Bristow et al, 2002), allow the user to call up web-based information about locations of interest as they approach specifically identified or named instances. A simple example illustrates a user following a campus map as they navigate the area, calling up information about various buildings as they approach each of these in turn. This type of scenario shows a potential application of such technology in the support of fieldwork within a number of disciplines. Although these examples may currently offer more benefit for lifelong learners, who have already developed the appropriate skills for learning, rather than being integrated into more formal learning scenarios, this does not preclude future applications along these lines.

Implications of Changing Content: Impact on Institutional Markets

As the various aspects of learning are teased out from formerly discrete institutional areas; which include institutional record handling, learner support, lifelong learning skills and career pathways, assessment function and technique, as well as actual taught and other course provision; the complexity of the nature of 'content', and how it is viewed, becomes apparent. As content becomes aligned across many new areas that belong to the process of reintegrating the institutional core business, it begins to devolve. This can be turned to advantage when considering that, on the learning side, there are more forms of delivery available than ever before. This in turn can support the increase in the types of learning opportunity or scenario we wish to present.

Risk management in the delivery of learning

The devolution of content may appear at first to increase the complexity of how delivery can assist learning, but there is now a better understanding of what each type of delivery is strong at achieving (Hodgson, 1984; Bacon, 1996; Bacon & Swithenby 1996). This offers some attractive options for improving the learning opportunities available to existing student groups (including through more effective use of contact time), and to reach markets and students not easily assisted without technological communication and delivery backing (Li & Zhang, 2003; Crowley, 2000). Considering the first, for example, there are now ways in which the difficult aspects of a discipline can be introduced, so students have a better conceptual base from which to begin. This is particularly well demonstrated by the use of simulation and modelling within subjects that deal with material that is difficult to visualise (Song & Lee, 2002; Pasqualotti & Freitas, 2002) and for areas that particularly benefit from the introduction of real-life or collaborative scenarios (Hughes et al, 2001; Ruiz et al, 2002). Here, activities can begin with opportunities to visualise or experience a concept, or such opportunities can be ordered in such a way as to illustrate to students their evolving change in perception around the topic, at various stages of their learning.

However, changes in modes of delivery allow, and often require, the core business of an institution to be reassessed, not only in terms of realistic market expansion and retention, but in terms of ability to underwrite the risk associated with high-quality online developments. Sustainable developments must align effectively with an institution's capacity to service and retain both its existing markets and to attract new business in areas that are either sufficiently high-profile as to be worth the investment, or which are capable of generating sufficient income to sustain the course or programme for which they have been created. Associating high-investment developments with a strong need base allows effective targeting of resource to strategic areas, and a coherent cross-function approach to be taken. This in turn reduces the risks undertaken by an institution in underwriting the production of online provision and increases the chances of the end student experience being positive.

Much of the major risk involved in the development of a good standard of online learning and production of generically available software and systems that support learning has been underwritten by national funding bodies with institutional support and this work has been largely project-based (See Chapter 4). However, as institutions begin (and are required) to take over this responsibility as part of mainstream provision, effective institutional management of change processes and content processes adopts greater significance in view of the increased level of risk this implies for the business. This is likely to include an introduction of some type of project management structure to both relatively small as well as large-scale developments of content, even at academic level. This may in fact be so " light a touch" as to be almost imperceptible, if integrated effectively into regular departmental and faculty structures, provided that it can ensure that effective development of courses and delivery of student obligations, as an institution, are met. Given the extent of resource commitment required to create this type of electronic provision (Bacsich et al, 1999), failure to take account of risks through inadequate management, is arguably equally as damaging to the long-term future of an institution, as avoiding the demand for new course structure, design and student support that is now expected by new intakes.

Recognition of the risks involved for institutions, in underwriting electronic provision, is demonstrated by new moves to include provision within employment contracts of staff, for circumstances arising due to key contributors changing their employment (HEFCE, 2003). Such moves recognise the rights of an individual to their own academic material produced for such purposes, and generally allow the transferability of content by authors, but retain the institutional right to continue running the course and to make any relevant adaptation and updates. Generally freedom is allowed for the staff member to continue to run a similar course from their new institution, often on the understanding that this is unlikely to be identical (many courses being team produced). There is also usually an assumption that it will continue to evolve with that staff member in their new institution, as will the course 'left behind' and thus both will fill separate niches in the longer term. Equally, a member of staff no longer involved in a course that bears their name, may request removal of any association with themselves, such as in cases where they believe the material represented to no longer be in date. Contractual arrangements for non-contributing course staff, such as maintenance and technical development, fall under different constraints and so are likely to handled separately. It is conceivable in some circumstances that an institution may require a development contract, covering many of these issues, to be signed by all relevant staff before development and presentation of such a course is institutionally approved.

Models for change

In view of the significance of institutional change required to successfully operate in the area of online provision, Koper (cited in deVries 2003) has developed the 'Learning Networks' model of "connecting people, organizations, autonomous agents and learning resources to establish the emergence of effective lifelong learning". This aims to develop a coherent set of sub-models and new learning technologies to create and support a new approach to lifelong learning. The development of this model follows the success of the Open Universiteit Nederland (OUNL), in developing learning technologies for educational modelling, competency-based approaches within electronic learning environments, and in tools for authoring and deployment of learning resources for e-learning. EML, the Educational Modelling Language developed and implemented by the OUNL, has since been adopted as the basis for the IMS Learning Design Specification.

The Learning Networks model maps examples of organisational structure and changes in the ways of working as opportunities are opened up due to developments in technology that offer advantages to learning organisations [2]. Initial discussions within a combined Dutch-UK study group that I was a member of focused on the 'abstract node model' proposed by Koper. In particular we discussed the implication that a rollout model for institutional change is an abstract concept. In this context, application of the model logically implied large risks for all parties and the institution itself, alongside requiring an overwhelmingly large resource input in one operation. In addition, the underlying assumption was included that the outcome of this will be that high costs will be deferred to the students and be (falsely) re-presented as higher efficiency for the institution. With its infinite cost and resourcing implications, this made the model, in this particular context, appear untenable to us.

However, the weaknesses in the model are not as immediately apparent as this. In fact, the model itself appears strong and tenable, at least for conditions where good institutional management policy is invoked to avoid this conflict situation. Targeting policy needs and requirements must be based on a re-evaluation of the market position and whole business of the institution. This needs to be done through identifying viable extensions and 'gaps' for existing provision. Then resource allocation can be aligned (people, tools, cost) to the new provision (risk) and application of the new policy. The process can then be controlled along the already-identified tension lines in a managed way, rather than spiralling out of control as has been suggested. Subsequent generic manageable outcomes and benefits can be rolled out over the current student/course and business model and new developments so stay within a sustainable and scalable resource model.

Implications of Changing Content: Impact on Individual Enthusiasts

Changes to content within a well-managed institutional position offer some immediate benefits to those staff who are individually responsible for supplying the knowledge base for any given online course. Development support is more widely available, often in a more comprehensive yet standardised form; 'rights' conditions for individual's work within a team are becoming more clearly defined and the element of risk to the individual in introducing new material and concepts can be reduced.

A new freedom

The current trend is to remove the development loading from academics, and to offer services and reuseable material on which they can easily draw. This allows them to concentrate their time on producing good learning foci for their students with support from instructional designers, resource specialists, and learning technologists. This may leave many of the original multi-skilled entrepreneurs feeling constrained in what they can achieve as they become freed up in this way, due to mainstreaming of course production and technology-use within learning – a field which they have in effect pioneered.

In the past, the material and delivery for a course largely rested with the tutor who specialised in that area, or self-contained core courses existed that were based on published texts and problems and practical or fieldwork, that circulated within a department using its own equipment and teaching facilities. From this environment, individual practitioners were fairly free to experiment with new modes and methods of learning presentation and delivery, and with some provisos, to handle the whole process them-selves. This offered a rich learning opportunity for such entrepreneurs, from which a great deal of today's experience and good practice is derived. Now course development has become a multi-disciplined activity involving many specialist professionals brought together as a team. In this environment, roles become redefined, with the result that some of the activities, from which these types of individuals obtained their greatest learning, are now less open to them within the new institutional environment. Team skills also become more important, as many specialist groups come together to initially analyse the needs for each intervention, and determine the level at which it requires build and implementation support. In practice, many of these entrepreneurs have re-tained their variety of skills either by moving into related areas generated by these new developments, or have applied their knowledge to effective use in extending main-stream handling of longstanding areas of learning.

New roles for tutors and academics

Although these changes may look like greater restrictions on the role of the tutor or academic, it still leaves the individual practitioner with a great deal of freedom in de-termining the learning objectives, structure and content of any course in which they are involved in developing. And as each specialist learning area becomes more intense, this apparent narrowing of responsibility in fact keeps the loading manageable. In some cases, the practitioner role will also include decisions as to what sections of the course may be presented in which forms. This approach remains in contrast to situations where course content tends to be defined by an external examination syllabus, and where some or all of the materials and course structure are supplied in a pre-defined form to the tutor and student (pre-tailored centrally to specific context). Such structures can often be ex-tremely rigid; the extent to which frequently depends on the flexibility that can be ac-commodated within implementation of the learning scenarios. Similarly, where com-plex re-usable resources are available, lecturers and tutors often need to 'tailor' these in some way to suit their own purposes. Use of such resources is recognised as requiring some fluidity, given that the specific contexts where learning resources are used will often differ within apparently similar courses, even where curricula are externally de-fined. This situation does not arise to the same extent for individual 'classroom-produced' resources, which frequently involve less complex production techniques and forms of delivery, of types that are more commonly used within delivery-limited situa-tions.

New approaches for students

The approach to inducting students into courses has also changed to accommodate the learning process more, and to de-emphasise knowledge-based content and passive tech-niques for study. We frequently begin our teaching by challenging and questioning the learner right from the start. A course may be opened with a question, an activity, or a short quiz to test current knowledge, in contrast to the frequently passive lecture or text mode that has been adopted within many courses in the past (Hodgson, 1984). This serves not just to establish a student's initial knowledge, but to lay the ground for activ-ity during learning and induct the student to expect a process of constant challenges.

The "mould" is set from the start of the learning process, and does not need to be changed and reformed part way, or invoked by a process of gradual familiarisation. Although the student can no longer leave thinking and review until it is convenient, in many online and flexible study cases, the trade-off is that they are now free to choose the time and place at which the activity will occur, so that they are best prepared for the intervention. These active approaches apply equally within face-to-face and technology-based environments for learning, reflecting an overall change in process within the sector irrespective of the medium of presentation. Through reflective practice, the onus to introduce these changes in styles to improve student learning is being added to the new role of the academic or tutor.

In the development of technology-based presentations, such changes in learning stimuli become apparent through the appearance of more strategically situated activities throughout module work. These activities are in place of text-based work, or a passive script start, followed by activity and/or assessment derived from a linear approach. Good educational designers have adopted this technique in various ways for many years. This approach has in part been encouraged by schemes such as the European Academic Software Awards (EASA). But with the advent of the Internet and use of the hyperlinked text-based resources, a lot of online development in the establishment period for these technologies within education has reverted back to a passive text-based information-assessment approach. This is in part owed to the relative ease of access of these techniques to individual pioneering academics, during the early years of the Internet and the Web, than software development had previously been. However, this legacy has left a fairly strong development curve to be followed in encouraging re-thinking of the actual learning achievements that educational use of the technology has offered, and has required some time to redevelop as an ethic across the sector.

Conclusions

Changes in individual and mainstream thinking on student learning, reflective practice, and preparation for equipping individuals to handle major changes with the way their world operates, have caused an evolution in the form and purpose of content. Evolution of content is rather like those organisations that are set up with the purpose of eliminating themselves. Once their core purpose is achieved, there is no longer a need for that function. A simple example might be a pressure group set up to change a particular aspect of the law, and its position once that change has been put in place. Yet similarly, those initially involved will often redevelop and move onto related areas that have evolved or been created from that original activation. Through its evolution, content will still be present but no longer recognised in that form. It will be an integration of resources, feedback, facilitation, and skill definition, backed by presentation media and the practical settings in which it is introduced. New issues will arise for learning from these new conceptions, and those of us with a prime interest in these areas will find new stimuli in what has arisen as a result of these changes.

References

Bacon, R.A., & Swithenby, S.J. (1996). A strategy for the integration of IT-led methods into physics -- the SToMP approach. *Computers in Education*, 26, 135-141.

Bacon, R. (1996). The effective use of computers in the teaching of physics. *Active Learning*, 4, 37-41.

Bacsich, P., Ash, C., Boniwell, K., & Kaplan, L. (1999). *The Costs of Networked Learning.* Sheffield Hallam University: Telematics in Education Research Group.

Bristow, H., Baber, C., Cross J. & Wooley, S. (2002). The χ^3 wearable computer and WECA PC. In S. Anastopoulou., M. Sharples & G. Vavoula (Eds.), *Proceedings of the European Workshop on Mobile and Contextual Learning*. Birmingham: The University of Birmingham.

Calverley, G., & Shephard, K. (2003). Assisting the Uptake of Online Resources: Why Good Learning Resources are not Enough. *Computers in Education. in press.*

Candy, P. (1988). On the attainment of subject-matter autonomy. In D. Baud (Ed*)* *Developing Student Autonomy in Learning* (pp59-76). London: Kogan Page.

Crowley, M. (2000). Startech - Learning Together: Video-conferencing in Geographically Remote Areas. In *Proceedings of NCTE Conference: Schools Integration Project Symposium* (pp42-46*)*. Dublin: National Centre for Technology in Education.

Gadzella, B.M., Ginther, D.W., & Bryant G.W. (1996). Teaching and Learning Critical Thinking Skills. *International Journal of Psychology,* 31 (3-4), 3247

Gilb, T. (1998). *Principles of Software Engineering Management*: Boston, MA: Addison-Wesley.

HEFCE (2003). *Intellectual Property Rights in e-Learning Programmes*. Report of the Working Group, Good Practice Guidance for Senior Managers.

Hodgson, V. (1984). Learning from Lectures. In F. Marton , D. Hounsell & N. Entwhistle (Eds.) *The Experience of Learning.* Edinburgh: Scottish Academic Press.

Hughes, C.E., Moshell, J.M., Reed, D., Chase, D.Z, & Chase, A.F. (2001). The Caracol Time Travel Project. *Journal of Visualisation and Computer Animation*, 12, 4, 203-214

Knight, P. (2002). A Systematic Approach to Professional Development: Learning as Practice. *Teaching and Teacher Education* 18, 3 229-241.

Kreber, C. (1998). The relationships between self-directed learning, critical thinking, and psychological type, and some implications for teaching in Higher Education. *Studies in Higher Education*, 23,1, 71-86.

Li, C., & Zhang, X. (2003). ' IP/multicast video conferencing for distance learning'. *Journal- Tsinghua University,* 43,1, 129-131.

Malhotra, Y. (1996). *Enterprise Architecture: An Overview* [On-line] Available: http://www.brint.com/papers/enterarch.htm

McDowell, L.(2002). Electronic information resources in undergraduate education: an exploratory study of opportunities for student learning and independence. *British Journal of Educational Technology*, 33,3, 255-266.

McHardy, P. & Allan, T. (2000). Closing the gap between what industry needs and HE provides. *Education and Training*, 42,9,496-508.

McHardy, P., & Henderson, S. (1994). Management Indecision: Using group-work in teaching creativity. In R.Gregory & L. Thurely (Eds.) *Using Group-based Learning in Higher Education*. London: Kogan Page.

Page, S., & Meerabeau, L. (2000). Achieving Change Through Reflective Practice: Closing the Loop. *Nurse Education Today*, 20,5, 365-372.

Pasqualotti, A., & Freitas, C.M.D. (2002). MAT(3D): A virtual reality modeling language environment for the teaching and learning of mathematics. *Cyberpsychology & Behaviour*, 5,5,409-422.

Rader, H. (1995). Information Literacy and the Undergraduate Curriculum. *Library Trends*, 44,2, 270-278.

Reynolds, M. (1998). Reflection and Critical Reflection in Management Learning., *Management Learning*, 29,2,183-200.

de Ruijter, C., & Maas, J. (2003) *Site Visit to Industrial Design Engineering*, NL-SURF Exchange 7-{On-Line}. Available: http://www.brookes.ac.uk/research/odl/alt-nl03/NL03_monday.htm

Ruiz, I.L., Espinoza, E.L., Garcia, G.C., & Gomez-Nieto, M.A. (2002). Computer-assisted learning of chemical experiments through a 3D virtual lab. In P.M.A Sloot, C.J.K Tan, J. Dongarra & A.G. Hoekstra (Eds). *Proceedings of ICCS 2002: Lecture Notes in Computer Science* (pp. 704-712). Heidelberg: Springer-Verlag

Sharples, M., Rudman, P.D., & Baber, C. (2002). Supporting Learning Conversations using Personal Technologies. In S. Anastopoulou, M. Sharples & G. Vavoula (Eds.). *Proceedings of the European Workshop on Mobile and Contextual Learning*. Birmingham: The University of Birmingham.

Song, K.S., & Lee, W.Y..(2002).. A Virtual Reality Application for Geometry Classes. *Journal of Computer Assisted Learning*, 18, 2,149-156.

Taylor, J., Peake, L., Philip, D., & Robertshaw, S. (2002). Location Activated Nomadic Discovery (LAND). In S. Anastopoulou, M. Sharples & G. Vavoula (Eds.), *Proceedings of the European Workshop on Mobile and Contextual Learning*. Birmingham: The University of Birmingham.

Tiwana, A., & Bush, A. (2001). A social exchange architecture for distributed Web communities. *Journal of Knowledge Management* 5, 3,242-248.

de Vries, F. (2003). *New Learning Technology Development Programme, Learning Networks*. [On-Line]. Available:
http://learningnetworks.org/forums/showthread.php?s=&threadid=97

Notes

[1] CETIS Educational Content Special Interest Group presentations: http://www.cetis.ac.uk/

[2] Learning Networks: http://learningnetworks.org/

3

Learning Objects, Pedagogy and Reuse

Tom Boyle and John Cook

The issue of the effective reuse of learning resources has become the focus of considerable interest. There are major economic and pedagogical reasons for capturing quality resources in a form that facilitates reuse in a variety of settings. The issue of reuse is closely entwined with the concept of 'learning objects,' which is used to refer to the basic reusable unit of learning. There has been a major international effort in developing standards to facilitate the reuse of resources. Duval et al (2003) argue that this work should increase the effectiveness of learning by making content more readily available, reducing the cost of producing quality content and allowing content to be more easily shared.

This work has made significant progress. However, in order to make the specifications and standards as widely applicable as possible, the concept of what constitutes a learning object was left at a very general level. The concept of learning objects, however, turns out to be problematic. McGee (2003) expressed the problem bluntly: "the standards have come before the thing itself has happened". There is a clear need to clarify the concept of what is the basic unit of reuse.

This chapter begins by examining what we mean by 'learning objects' and 'reusability'. The examination points to the need for a model to understand these issues rather than a debate over the exact meaning of the phrase 'learning object'. The discussion then focuses specifically on the design of learning objects that are both reusable and pedagogically effective. It outlines principles that facilitate reuse, and argues that these can be combined with rich pedagogical techniques. The use of such learning objects in a large case study is reviewed and ways in which such objects can be placed in electronic libraries, consonant with the international standards that support learning object portability are explored. Finally, issues of support for staff who are reusing learning objects are examined. The key theme throughout the discussion, is on the pedagogical significance of learning objects.

What Exactly Are Learning Objects?

The first approach is to view a learning object as the entity around which standards for describing and importing learning resources could be built. The concept is therefore described in the broadest possible terms. For example, the IEEE (March 2002) standard for learning object metadata defines learning objects as:

> [...] any entity, digital or non-digital, that may be used for learning, education or training.

However, the concept of reuse points to the idea of basic building blocks which can be treated as independent entities and combined in news contexts of use. Most of the attempts to clarify the concept of a learning object have been concerned with trying to identify and describe this basic unit of reuse. Thus, Polsani (2003) defines learning objects as:

> [...] an independent and self-standing unit of learning content that is predisposed to reuse in multiple educational contexts".

The key characteristics that emerge are granularity and reusability. Two key issues thus emerge: clarifying the nature of this basic unit of reuse, and clarifying exactly what we mean by reuse. Both these issues were discussed in some depth at the Learning Objects Symposium held before the Ed Media conference in Hawaii in June 2003. Many world experts in learning objects and standardization attended this event. The symposium also included experts from quite different pedagogical traditions (instructivist and constructivist). In this symposium, Eric Duval argued that it is more important and productive to try to find a model for understanding learning objects rather than producing a definition. Wayne Hodgins echoed this position by arguing that what we need is a principled base model rather than a definition. Within this model it may then be possible to articulate different types of learning object. The approach followed in this chapter is that it is more important to clarify the model rather than argue about precise definitions.

Learning Objects: Content and Pedagogical Value

The primary approach to reusability is to separate content and context in order to make the content reusable in different contexts of use. Learning objects are thus viewed as chunks of content. Wiley extends this position by arguing that here is a direct inverse relationship between size and reusability (Wiley, 2003; Wiley et al. 2003). To maximise reusability these chunks of content should be as small as possible. He argues in particular, that the "fit" between the internal context of the learning object and the external context of use determines the possibility for reuse. The less specific the internal pedagogical context the greater potential there is for reuse. He thus proposes that learning objects should consist of small chunks of content.

The problem with this argument is that if it is extended too far you get a 'reductio ad absurdum'. The smallest and most reusable unit is the pixel. However, it does not seem a strong candidate for a learning object. There clearly is a countervailing consideration: how pedagogically useful is the unit?

Boyle (2003) proposes a different basis for demarcating basic learning objects. It is the identification of minimal meaningful educational objectives that provides the basis for learning objects. The learning objects he proposes are "learning microcontexts" structured to achieve these educational objectives. They explicitly consist of the content plus the pedagogical processes required to achieve the educational objective. This approach has been applied to developing learning objects for programming (Bradley & Boyle, 2003; Bradley, Boyle & Haynes, 2003). The learning objects are based on minimum basic learning units, which it does not seem pedagogically sensible to subdivide. Thus if there are three types of loops in a programming language then one learning object is produced for each loop type. Boyle (2003) provides authoring principles to ensure that the objects are designed for maximum reuse.

The differences in opinion between Wiley and Boyle point to a key issue that was discussed in depth at the Ed Media symposium (Duval et al, 2003). What is the primary criterion for learning object demarcation: reusability or pedagogical value? One position is that learning objects are content chunks; the pedagogical value is created in the context of use. A catalogue of pictures, for example, could be a useful educational resource even though each picture contains no explicit pedagogical material. The opposing position is that these are 'information objects'; learning objects should contain intrinsic pedagogical value (e.g. pictures with pedagogically oriented commentary).

As Duval argues, we need a model, not contestations over a phrase. These two conceptions seem to point to different layers of reuse. The pedagogical objects, that contain explicit pedagogical processes, operate at a higher level than the more basic information objects. The pedagogical objects will often (maybe always) contain information objects. We need to clarify the relationship between the two rather than engage in sterile debate over nomenclature. This exploration seems to point to the idea of several layers of reuse. The lower layers contain less pedagogical value but are more flexible in terms of use in a wide variety of contexts. The higher layers have more limited scope for reuse but have higher intrinsic pedagogical value. If we have this range of choice available then the local tutor can decide on the resources most useful for their particular needs.

Conceptions of Reuse

People agree that reusability is a central criterion for learning objects, but they have different conceptions of what they mean by 'reusability'. There are stronger and weaker versions of how people conceive of reusability. The 'weaker' view is concerned solely with the creation of electronic libraries. The items in these libraries may vary in size and type. The key issue here is being able to describe the items using metadata. Users may then search the library to locate items that meet their needs. Reuse in this scenario depends on being able to locate something in a library. How that item is used and combined with other items is left to the local tutor. The pedagogical value is supplied by the local tutors in the teaching and learning context they create. This version places very little restrictions on what may be placed in the library.

The 'stronger' version of reusability, views learning objects as basic units selected or designed to have properties that facilitate their combination into higher order learning units. The crude analogy is with the Lego brick. The key point about a Lego brick is that it is designed to be a basic unit out of which higher order structures may be built. So the stronger view aims to designate as learning objects only those basic units that have properties that specifically support (re)combination into higher order pedagogical units. If we combine the different views on reusability with the different conceptions of 'basic' learning objects we finish up with three positions, all of which are current:

1. A learning object is "any entity ... that may be used for learning, education or training" that has appropriate metadata added to aid retrieval from an electronic library, e.g. lecture notes, diagrams, pictures etc. This represents the broad pragmatic position with many 'learning object' repositories (e.g. Koppi & Hodgson, 2001; Koppi & Lavitt, 2003; Hanley, 2003; MERLOT [1]);
2. Learning objects consist of basic content chunks optimised for recombination into higher order structures where pedagogical process is added;
3. Learning objects are basic 'learning micro-contexts' explicitly designed for flexible (re)combination into higher order pedagogical structures. These micro-contexts are organised around basic learning objectives.

We can deal with these different positions at a pragmatic or a conceptual level. At a pragmatic level we can continue to build and use repositories to hold 'learning' objects of these different types. The standards developed for learning object packaging and description support this pragmatic use. At the conceptual level there is a need for clarification. This returns us to the central argument, that there is a need to build a model in which these different conceptions can be placed in an articulated relationship to each other. One important concept in articulating this relationship is that of different layers of reuse. The present arguments may turn out, to some extent, to be about the different levels of reuse. Rather than being polar opposites, these views may turn out to have a complementary relationship to each other.

Learning Objects and Constructive Learning

The concept of learning objects as chunks of reusable content is associated primarily with instructivist approaches focused mainly on individualised training. Wiley (2003:1) criticises the pedagogy associated with this approach as outdated. He argues that: "while they harmonize well with 1980s learning research, the assumptions of current learning object approaches frequently contradicts recent research on learning". He also criticises the conception of "info-capsules that transfer inert knowledge" (Wiley 2003: 3). If learning objects are to be of serious interest to those concerned with education then it must be demonstrated that they are consonant with modern constructivist conceptions of learning. The remainder of this chapter concentrates on the issues of the design and evaluation of reusable learning objects that support active student-centred learning. The focus is thus on what has been termed earlier in the chapter as 'pedagogical objects' rather than 'information objects'.

Structural Design of Learning Objects

The design of learning objects involves two main considerations: structural design to ensure reusability and pedagogical design to ensure effectiveness. The two are ultimately connected. The purpose of the structural design principles is to facilitate the reuse and flexible recombination of the learning objects. This facilitates the distribution of the development effort across several sites. Good pedagogical design and development is an intensive business. By distributing this development effort across many contributing sites a critical mass of high quality learning objects may be achieved.

The main structural principles are 'cohesion' and 'decoupling'. Cohesion means that each object should do one thing and one thing only. Each learning object should be organised to meet one clear learning goal or objective. The content and processes inside the learning object should concern that objective and that objective only. This discipline can sometimes be difficult for tutors. There is a natural tendency to elaborate. The consequence, however, can be cognitive overload for students. An example is provided by a workshop group we participated in, that was developing a learning object for a particular statistical concept. The group found it difficult to remain focused on one clear learning concept. They finished up discussing at least four different statistical concepts. The danger, however, is to overload students with premature complexity. The principle of cohesion is proposed primarily as a structural feature to ensure the identity of the learning object. However, it may also have benefits in pedagogically intense subjects such as statistics in ensuring that tutors consider carefully the cognitive load they place on students.

The principle of decoupling is more accurately expressed as minimised coupling. The learning objects should have minimal bindings to other units. This is necessary to ensure maximum flexibility to reuse the object. There are two main types of binding that need to be avoided: navigational and content bindings. The navigational ('go to') bindings to any other units should be minimised and controlled. A typical hypertext document with multiple embedded links is thus a poor candidate for a learning object. It is very difficult to move this unit for reuse in a different context because it is too tied into the original context of use. Equally the content in the object should not refer to content in another source so that the content is unclear without the original source being present (e.g. 'extending the example from the previous chapter...').

Learning objects that are cohesive and decoupled provide maximum scope for reuse and recombination into higher order pedagogic structures. However, it might be argued that such focused learning objects might be pedagogically impoverished. Learning objects are only really useful if they achieve worthwhile pedagogical objectives.

Pedagogical Design of Learning Objects

The pedagogical implications of learning objects impact at two levels. At the first level the objects themselves must succeed in helping learners attain specific learning goals. While this is worthwhile, a simple list of such objects would ultimately be of limited overall value. The learning objects should also be designed to facilitate their combination or embedding in higher order pedagogical structures such as lessons, classes and courses. This is why structural principles are important. The immediate challenge, however, is to show that these cohesive objects can be pedagogically rich.

Rather than deal with this topic in the abstract, we will examine how pedagogical richness was incorporated in a series of learning objects that were developed to help deal with an urgent and serious educational problem. There is a well-documented crisis in the teaching and learning of introductory programming at university level (Jenkins, 2002; Jenkins & Davey 2001). Failure rates in programming are very high, and this is a widespread problem. There are problems of complexity and abstraction that can lead to alienation and disengagement often at a very early stage. The learning objects we developed were designed to help students deal with the learning challenges produced by these features of the domain. To do this we believed that we had to build in, as far as possible, constructivist learning techniques into the learning objects. As constructivist techniques are more normally associated with higher order domains this was an interesting design challenge. It should be noted, however, that Piagetian notions of constructivism would imply that constructive learning should be very pertinent at this level (Piaget, 1970).

In order to provide an integrated illustration of the potential for pedagogical richness the design of one learning object will be followed through. The object dealt with the basic loop construct; the 'while loop'.' One of the first aims was to engage the student and help them to relate this 'alien' concept to familiar everyday experience. Figure 1 shows the second screen in the 'while loop' learning object (the first screen simply introduces the object).

Our experience is that students can become disengaged from programming at a very early stage. Programming is failure critical. The concepts seem abstract and removed for many students. The programming environment is complex. It is easy to fail and repeated failure is highly discouraging. The screen in Figure 1 relates the example of a while loop to a non-threatening everyday example. It illustrates that repeating actions until a goal is achieved is a familiar problem solving technique. The screen is 'fun' and helps engage the student with the new domain.

A significant problem in many educational domains is aiding the student to construct a mental model of the topic. Figure 2 illustrates how visualization was used to help students construct an understanding of the code. In this screen the loop is used to move an object across the screen. The students can step through the loop at their own pace. Each step in the code has a visual effect on the screen (each repetition of the loop moves the submarine 50 pixels down the screen). The aim is to provide the student with the experience from which they can construct a mental model of how the loop works. The actual Java code is used, rather than a pseudo-English description, because the aim is to relate this authentic representation to the effects it produces. The use of visualization as a means of bridging from familiar to new abstract knowledge has been strongly

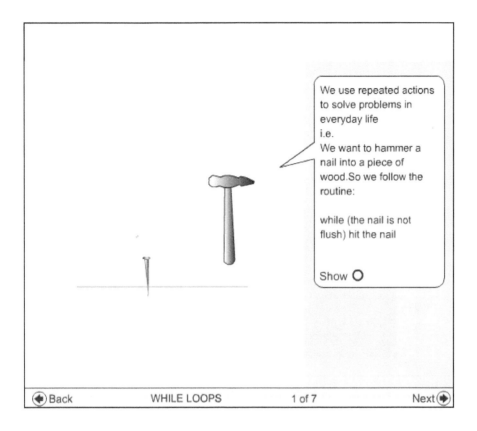

Figure 1. Encouraging student engagement by relating the concept to a familiar, everyday example.

advocated by theorists such as Papert (1980, 1993). This technique was used repeatedly across the objects developed. Thus in learning about arrays, for example, the operation of the surface code is related to a visual model of the array in computer memory.

Constructivism emphasises active learning. An important technique for helping learners deal with complex domain is 'scaffolding'. This technique has its origins in Vygotsky's ideas of the zone of proximal development (Vygotsky, 1962), and was extensively investigated by psychologists such as Bruner (1975). The concept is particularly pertinent to a complex domain such as learning to program. The basic concept is that the learners are provided with transitional support in learning a task that would be too difficult for the learner to handle without support.

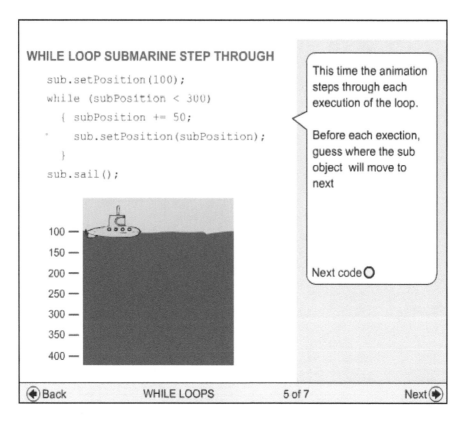

Figure 2. Use of visualization to facilitate the construction of understanding.

Interactive visualization is an important technique for assisting learners to build a mental model. However, it is important that the learners move to actively constructing the code. In the full development environment there are a myriad ways to fail and feedback can be potentially confusing (e.g. a semicolon in the wrong place in many programming language can produce an incomprehensible list of error messages from the compiler). Figure 3 illustrates how scaffolding was incorporated in the learning objects to help the learner make the transition to the full development environment. The learners construct code under supportive controlled conditions. The 'game' asks users to select the code lines to build up the five lines of code for the loop. Each selected line leads to feedback. When the code is complete the action initiated by the code, moving the horse objects across the screen, is illustrated.

These learning objects are quite compact. They each focus on only one programming construct. However, this curriculum focus does not prevent the deployment of rich, constructivist pedagogical techniques. In fact, it was regarded as essential that such techniques be incorporated in these objects, as the standard didactic textbooks were not working. The key message is that 'small' learning objects can and should incorporate rich constructivist techniques that encourage active learning (Examples of the learning objects including the 'while…loop' object can be accessed from the LTRI web site [2]).

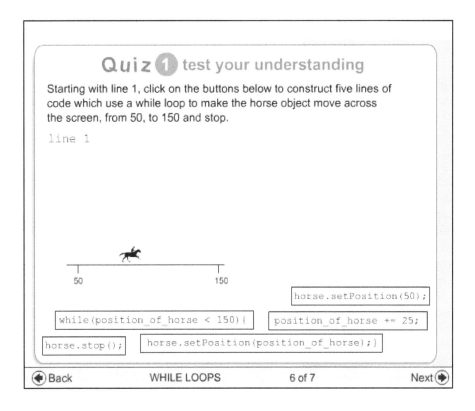

Figure 3. Use of scaffolding.

The Deployment and use of Learning Objects

There are two issues in embedding the objects into a particular course. At a technical level how are the objects accesses by the students from the standard VLE they are using? At a conceptual level what is the role and place of the learning objects in the overall course or module? In our project we used a simple practical technique for linking the learning objects to the Virtual Learning Environment (VLE) the students were using (in this case, WebCT). The learning object existed totally outside the WebCT environment; they were mounted on a separate server. The objects were simply dynamically loaded in at runtime. To the students the learning objects were accesses seamlessly from WebCT and probably seemed part of the WebCT environment. However, the objects were independent and could be as easily linked into by any other VLE.

The overall aim of the project was to improve success rates in programming. The learning objects were used as part of a new blended learning environment. This blend included changes in the offline organization of the course, the development of a substantial support environment in WebCT and the introduction of both text based and multimedia learning objects. Learning objects are not meant to provide a complete learning environment. They will normally need to be integrated into higher order course structures. The technique in this case was to include them as components in a particular blended learning environment. However, the flexibility of learning objects means that they can be reused in different ways. They could be used as components of a more automated system where 'intelligent' software uses learner profiles to generate particular learning paths for students. The articulation and separation of the learning objects from a courseware layer provides great flexibility in how the learning objects are selected and used at this higher layer.

Distributed Reuse for a Wider Community of Practice

Good, pedagogically effective learning objects take time and effort to produce. If the development effort can be distributed over institutions and individuals, who share the results, there is a basis for creating a critical mass of high quality learning objects. This notion of the distributed reuse of learning objects is often referred to as the "learning object economy", where widely accessible lifelong learning opportunities are made available through an open exchange of learning objects. However, developing the learning object economy is proving to be more complex than was initially envisaged.

There are many issues emerging from attempts to move towards a learning object economy. One key issue is the creation of repositories that would hold these learning objects. The learning objects would be stored in these repositories, and retrieved and downloaded when required by tutors. A local Virtual Learning Environment (VLE) should be able to recognise and handle the learning object reducing the effort required by the local tutor in incorporating it in their course. International organizations such as the IMS Global Learning Consortium (IMS, 2003) have been very active in developing the specifications that would lead to such standards. The two most relevant areas for this discussion are content packaging and metadata.

The basic idea of content packaging is to create a container or software wrapper into which the learning object is placed. There is a standard structure, which enables the package to be passed across different software systems and be recognised and manipulated by these systems (e.g. different VLEs). The key extra element is the manifest file. This contains metadata that describes the learning object, a description of the structure of the object, and pointers to the files that implement the elements of the formal organization. A prime function of the standard for content packaging is to enable interoperability across different software systems. Thus a learning object developed in one VLE could be packed and exported to another VLE, which would recognise and accept it.

Content packaging remains a specification proposed by the IMS Global Learning Consortium. As an evolving specification it is open to change. This makes it difficult for tool producers to develop stable tools that will handle the content packaging. The most recent version of content packaging is version 1.3.1. For those wanting to implement content packaging we would currently advise seeking technical advice. CETIS provide up to date information on the development of international standards and specifications [3]. The X4L JISC programme is developing a set of tools to deal with issues such as content packaging [4]. Content packaging is not essential in setting up learning object repositories but it does offer the promise of seamless interoperability between different systems. Many repositories have been set up which provide a database that holds (the metadata) descriptions of the learning objects. The objects are then located by pointers to the original locations of the objects.

The National Learning Network (NLN) is a good example of a large UK learning object repository initiative that was initially established to provide further education (FE) and sixth form colleges with online learning resources. However, the programme of work is now being rolled out to the wider post-16 education sector. The initiative began in 1999 and to date has drawn in £156 million of UK government investment over a five-year period. In terms of the granularity of NLN learning objects, they were originally intended to provide 20 minutes of learning. The NLN home page gives users what is claimed to be "a 'one-stop shop' for information and support on downloading and e-learning materials that are now available free to FE colleges through the NLN partnership programme" [5].

Beetham, Taylor and Twining (2001) argue that the main barriers to reuse appear to be cultural at both individual and institutional levels. Academics require considerable support in moving from the model of the individual craftsperson towards team based collaborative development and reuse within a community of practice. The NLN web site provides some support for this move. The tutor area of the NLN web site provides a 'community area' where practitioners can share their experiences and good practice with each other, and a forum for exchanging ideas and exploring how to get more from the materials. Providing the facilities to build up a community of practice around the use of learning objects seems crucial if they are to be adopted (Beetham, Taylor & Twining, 2001). The NLN appears to have begun to move along this road.

FERL, is another UK Government funded information service and learning object repository aimed at the post-16 sector. Unlike the NLN, which makes use of outsourced agencies to develop learning objects, FERL relies upon practitioners themselves to provide learning objects, which are typically smaller than the NLN learning objects. This fact reinforces the point that perceptions about the nature and size of learning objects differ. In order to enhance the use of FERL learning objects, lesson plans are provided that indicate how these objects should be used [6]. This seems like a useful attempt to support embedding learning objects into a community of practice.

Mayes points out strongly that "to be useful, learning objects will require the mediation of human (teacher) judgement, even though these judgements can operate in a very distributed way" (Mayes, 2003). What is needed is a partnership between researchers and practitioners in order to develop and deploy pedagogically enriched models for the authoring and reuse of pedagogical objects. This has not always happened in the past, and this has created problems in getting tutors to accept and use the metadata descriptive scheme implemented in the IEEE standard (Currier, 2003).

Conclusions

There are different views of what constitutes a learning object. The answer to these different views is not to argue over terminology, though this does need to be differentiated. As Duval et al (2003) argue, however, the way forward is to build a model of what we understand by reusable learning objects. This model should distinguish between different types of objects and hopefully clarify the relationship between them (perhaps using the concepts of different layers of reuse: Boyle, 2002; Boyle & Cook 2001).

There are also different models of what is meant by 'reuse'. The first model of reuse is the pragmatic one of setting up electronic libraries containing a wide variety of learning objects. The second broad approach is associated with the second more precise conception of learning objects as basic units of learning. The learning objects here are designated as potential building blocks of higher order pedagogical systems. One variant is to view learning objects as modular chunks of content (Wiley, 2003). Pedagogical value is added in the higher order use of these objects. An alternative approach is to view learning objects 'as micro-contexts' for learning, organised to achieve specific learning goals or objectives (Boyle, 2003). This approach enables rich pedagogical processes to be incorporated within the learning objects. The division here seems to be not an absolute difference but rather a differentiation between 'information objects' and 'pedagogical objects', which reside at different layers of reuse.

Learning objects are not just about resource development and reuse. To achieve the full potential of reusing quality resources we need to tackle the cultural barriers inherent in traditional structures and approaches to education. There is a need for cultural change, especially in higher education. We need to move away from the model of the individual craftspersons all reinventing the wheel towards team-based collaborative development and reuse of high quality resources. There is a need for institutions to facilitate this co-operation and avoid institutional barriers to progress. The ultimate and considerable challenge is to create vibrant communities of practice for the creation and reuse of high quality educational resources.

References

Beetham, H., Taylor, J., & Twining, P. (2001). *SoURCE Evaluation Report*. [On-Line]. Available: http://www.eres.ac.uk/source/docs/met-del-2.pdf.

Boyle, T. (2002). Towards a theoretical base for educational multimedia design. *Journal of Interactive Media in Education*. [On-Line]. Available: http://www-jime.open.ac.uk/2001/boyle/boyle.html

Boyle, T. (2003). Design principles for authoring dynamic, reusable learning objects. *Australian Journal of Educational Technology*, 19,1, 46-58.

Boyle, T., & Cook, J. (2001). Towards a pedagogically sound basis for learning object portability and re-use. In G. Kennedy., M. Keppell., C. McNaught & T. Petrovic (Eds.) Meeting at the Crossroads. *Proceedings of the 18th Annual Conference of the Australian Society for Computers in Learning in Tertiary Education* (pp. 101-109). Melbourne: Biomedical Multimedia Unit, The University of Melbourne. [On-Line] Available: http://www.medfac.unimelb.edu.au/ascilite2001/pdf/papers/boylet.pdf

Bradley, C. & Boyle, T. (2003). The development and deployment of multimedia learning objects. In E. Duval., W. Hodgkins., D. Rehak and R. Robson (Eds.) *Proceedings of Learning Objects 2003 Symposium: Lessons Learned, Questions Asked* (pp13-18) Association for the Advancement of Computing in Education. [On-Line] Available: http://www.cs.kuleuven.ac.be/~erikd/PRES/2003/LO2003/Bradley.pdf

Bradley, C., Boyle, T. & Haynes, R.(2003). Design and evaluation of multimedia learning objects. In D. Lasner and C. McNaught (Eds.) *Proceedings of Ed Media 2003, World Conference on Educational Multimedia, Hypermedia and Telecommunica-*

tions, Hawaii, June 2003 (pp.1239-1245). Association for the Advancement of Computing in Education.

Bruner, J. S. (1975). The ontogenesis of speech acts, *Journal of Child Language*, 2, 1-19.

Currier, S. (2003). *Quality Assurance for Digital Learning Object Repositories: Can Authors Create Good Metadata?* Research paper presented at ALT-C 2003, Sheffield.

Duval, E., Hodgkins W., Rehak D. & Robson, R. (2003). Introduction. In E. Duval., W. Hodgkins., D. Rehak and R. Robson (Eds.) *Proceedings of Learning Objects 2003 Symposium: Lessons Learned, Questions Asked* (pp13-18) Association for the Advancement of Computing in Education. [On-Line] Available [On-Line] Available: http://www.aace.org/conf/edmedia/symposium2003.htm

Hanley, G. (2003). MERLOT: Multimedia Online Resource for Learning and Online Teaching. In D. Lasner & C. McNaught (Eds.) *Proceedings of Ed Media 2003, World Conference on Educational Multimedia, Hypermedia and Telecommunications*, Hawaii, June 2003 (pp.1281-1284). Association for the Advancement of Computing in Education.

IEEE. (2002). *Draft Standard for Learning Object Metadata*. [On-Line]. Available: http://ltsc.ieee.org/doc/wg12/LOM_WD6_4.pdf .

IMS Global learning Consortium. (2001) IMS Content Packaging Specification. [On-Line]. Available: http://www.imsproject.org/content/packaging/index.cfm.

Koppi, T., & Hodgson, L. (2001). Universitas 21 Learning Resource Catalogue using IMS Metadata and a New Classification of Learning Objects. In *Proceedings of World Conference on Educational Multimedia, Hypermedia and Telecommunications 2001*(1), 998-1001. Association for the Advancement of Computing in Education. [Online]. Available: http://dl.aace.org/8646

Koppi, T., & Lavitt, N. (2003). *Institutional use of learning objects three years on: lessons learned and future directions*. Paper presented at Learning Objects 2003 Symposium Hawaii June 24. [On-line]. Available: http://www.cs.kuleuven.ac.be/~erikd/PRES/2003/LO2003/Koppi.pdf

Jenkins, T. (2002). On the difficulty of learning to program. In *Proceedings of 3rd annual conference of the LTSN-ICS, Loughborough, UK*. Loughborough: LTSN-ICS.

Jenkins, T., & Davy, J. (2001). Diversity and motivation in introductory programming. *Italics*, 1,1. [On-line]. Available:
http://www.ics.ltsn.ac.uk/pub/italics/issue1/tjenkins/003.html.

Mayes, T. (2003). Vision and Theoretical Perspectives: Introduction to Part 1. In Littlejohn, A. and Buckingham Shum, S. (Eds.) Reusing Online Resources (Special Issue) *Journal of Interactive Media in Education*. [On-line]. Available: http://www-jime.open.ac.uk/2003/1/

McGee, P. (2003). Remark made during the opening panel of the Learning Objects 2003 Symposium. Hawaii, June 24 2003.

Papert, S. (1980). *Mindstorms: children, computers and powerful ideas*. New York: Basic Books.

Papert, S. (1993). *The children's machine*. New York: Basic books.

Piaget, J. (1970) Piaget's Theory. In P. H. Mussen (Ed.) *Carmichael's manual of child psychology, 3rd Edn.* New York: John Wiley and Sons Inc.

Polsani, P.R. (2003). Use and abuse of reusable learning objects. *Journal of Digital Information*, 3, 4. [On-Line]. Available:
http://jodi.ecs.soton.ac.uk/Articles/v03/i04/Polsani/

Wiley, D. (2003). *Learning objects: difficulties and opportunities*. [On-line]. Available:
http://wiley.ed.usu.edu/docs/lo_do.pdf

Wiley, D., Padron, S., Brent, L., Dawson, D., Nelson, L., Barclay, M., & Wade, D. (2003). Using O_2 to overcome learning objects limitations. In E. Duval., W. Hodgkins., D. Rehak & R. Robson (Eds.) *Proceedings of Learning Objects 2003 Symposium: Lessons Learned, Questions Asked* (pp61-70) Association for the Advancement of Computing in Education. [On-line]. Available:
http://www.cs.kuleuven.ac.be/~erikd/PRES/2003/LO2003/Wiley.pdf

Vygotsky L. S. (1962). *Thought and language*. Cambridge: MIT Press.

Notes

[1] MERLOT (Multimedia Online Resource for Learning and Online Teaching):
http://www.merlot.org/

[2] Examples of Learning Objects: http://londonmet.ac.uk/ltri/learningobjects/

[3] Centre for Educational Interoperability Standards: http://www.cetis.ac.uk.

[4] X4L Exchange for Learning Programme:
http://www.jisc.ac.uk/index.cfm?name=programme_x4l

[5] National Learning Network: http://www.nln.ac.uk/materials/

[6] FERL: http://ferl.becta.org.uk/

Acknowledgements

The 'learning objects for programming project' has been supported by the LTSN Centre for Information and Computer Sciences (http://www.ics.ltsn.ac.uk). We would also like to acknowledge the work put in by the project team: Pete Chalk, Ray Jones, Ken Fisher, Claire Bradley, Richard Haynes at London Metropolitan University, and Poppy Pickard at Bolton Institute.

4

Experiences of National Projects in Embedding Learning Technology into Institutional Practices in Higher Education

Jacqueline Dempster and Frances Deepwell

Over the last decade, the UK higher education funding councils have funded a staggeringly large number of educational development projects under a range of teaching and learning programmes. Of these, a significant number aimed to develop and implement methods and tools specifically involving learning technologies. However, the sector has witnessed varying degrees of success in the way the innovative approaches explored by project work have been embraced, adopted and embedded. An evaluation of project success factors is therefore useful in order to inform decisions about educational innovation in the future.

A study funded by the LTSN Generic Centre was undertaken to draw out lessons learned from a number of national projects that had shown success in embedding new practices into institutional teaching and learning. Projects were funded by the UK higher education funding councils between 1998-2002 and were generic or interdisciplinary, technology-related educational development projects working at institutional and multi-institutional levels. This chapter describes the experiences of the selected projects in institutional embedding, highlights approaches that appear to work well and identifies potential areas where embedding might be enhanced at project, institutional and national levels. It outlines the operational contexts in which educational development work is located and provides a practical framework for planning and organising project activities to take account of these.

The Context of Institutional Implementation: an Evolving Landscape

Higher and further education environments at national and international levels have rapidly evolved over the last decade supporting a shift in the quality agenda from regulation to enhancement and an increase in national support for teaching and learning development. The stream of national funding initiatives, programmes and services has been an enviable part of UK educational history. In higher education, key developments are seen in quality assurance and quality enhancement agendas, in funding for national and institutional development work and subject support networks, in the professionalism of teaching and rewards for teaching excellence, and in research and evaluation of education as a scholarly pursuit. Alongside these changes, the staff themselves have increasingly formed associations, networks and accreditation pathways for their own continuing professional development. The Staff and Educational Development Association (SEDA), the Association for Learning Technology (ALT) and the Universities and Colleges Information Systems Association (UCISA) were all set up in 1993 and the Institute for Learning and Teaching in Higher Education was established in 1999. Further regional and special interest groups, project clusters and subject networks have emerged to facilitate sharing and discussion of specific aspects of practice in learning technology.

It is not surprising that over the last ten years that the learning technology environment has both informed these changes and been influenced by them. The rapid evolution of national programmes and support for teaching and learning development coupled with the emergence of institutional strategies for Information and Communication Technologies (ICT) has driven expansion of a vast composition of specialised staff working within the field of learning technology. For example, in library and information services, the last decade has witnessed national funding of over 500 projects involving at least 1000 staff. However, the disjointedness in the funding and continuation arrangements has had two major consequences for work within projects, firstly for the staff and secondly for the projects themselves. In terms of the human resource aspects of such arrangements, most staff who contribute to national funded work do so with little job security. The short-term nature of such projects has serious implications for the careers of staff who work on them as well as for the institutions in which they are employed. The results of two recent studies served to increase understanding about the issues surrounding the recruitment, development and retention of learning technology, library and information services project staff in UK higher education (Beetham, Jones & Gornall, 2001; Chems, 2002).

Both nationally and institutionally, moves to join up thinking about educational development across a number of increasingly overlapping areas of academic business are apparent. In 2003, the Teaching Quality Enhancement Committee (TQEC, 2003) recognised the need to address the "widespread perception that the arrangements for quality enhancement are complex and fragmented", and insufficiently "user-focused" and to interweave more tightly the various strands of support that had been separately established over the years. This review of future needs and support for quality enhancement is likely to bring about several changes over the next few years in the way that learning and teaching in UK Higher Education is organised, developed and supported.

Changing Practice Through National Educational Development Projects

Three significant national programmes (TLTP, FDTL and ScotCIT) have supported learning technology development in UK higher education over the past decade. Each of these programmes called for project proposals from individual higher education (HE) institutions or consortia in a competitive bidding process. The Teaching and Learning Technology Programme (TLTP) was set up in 1992 with an aim within the first two phases of 76 projects to encourage the higher education sector to work collaboratively and explore how new technologies could be exploited to improve and maintain quality within teaching and learning. A third phase of around 30 projects in 1998 concentrated more on implementation and embedding of materials within institutions. The Fund for the Development of Teaching and Learning (FDTL), now in its fourth phase, was established in 1995 and has so far supported 94 projects throughout Higher Education Funding Councils (HEFCE) funded institutions. The programme aims to stimulate quality enhancement in teaching and learning in higher education by encouraging the dissemination of good teaching and learning practice across the higher education sector. FDTL supports educational development more broadly than TLTP, though there are some projects exploring technology based approaches. In Scotland, the C&IT Programme of the Scottish Higher Education Funding Council (ScotCIT) ran between 1998 and 2001 and sought to establish appropriate use of ICT as part of normal working practice at Scottish Higher Education Institutions. It comprised 19 individual projects spanning four integrated strands: staff development, Web tools, intranets and infrastructure. A distinctive aim was to encourage the outcomes of funded projects to be applied outside the institutions directly involved in the development work. Most of the projects were therefore funded with the express purpose of embedding rather than a secondary objective of innovating and developing.

One factor that all of these projects have in common is their fixed term duration and therefore their considerable start-up and wind-down overheads. The short-term nature of nationally funded projects has equally serious implications in terms of the effectiveness of the projects in influencing practice in the medium and longer term. It is clear that changing teaching and learning practices, especially at institutional level, involves a complex set of processes occurring over an often considerable time span in order to gain momentum. Projects locate themselves within the complexities of institutional change management. They may also inherit the difficulties central academic development units themselves have in encouraging teaching development at both departmental and policy level. The operational context for national educational development work follows the quality enhancement agenda and a mission of academic development more broadly, for which a short-lived project is unlikely to be fully equipped.

Despite these issues, it is true to say that some projects have had more success than others in embedding new approaches into institutional teaching and learning practices. Some of the factors influencing such success are well known to educational developers across the sector, but have rarely been documented and are usually evidenced on the basis of individual projects rather than generic studies. On the ground, embedding might mean that projects require a sophisticated understanding of curriculum design and change processes. Successful embedding of project approaches is therefore likely to favour best a model of research and development far more than one of implementation (Dempster & Blackmore, 2002:131). For change to take place at institutional levels, the isolated practices of individual lecturers and students that a project may explore within and beyond its lifetime need to be evaluated and integrated with mainstream institutional practices. Taylor (1998:273) suggests that "the challenge is to move beyond innovation at the level of individual subject or organisational element to change at the institutional level, the reinvention of cultures" and that "there is a need to recognise that the challenge is not limited to the development of innovation, but extends to the institutionalism of the outcomes of the innovation." It is certainly evident that the two distinct but complementary 'quality' remits in an institution (quality assurance and quality enhancement) are often politically and operationally divided in terms of a university's organisational structures and processes. As with all academic development work, therefore, the institutional context within which these projects operate is crucial to their success in the longer term.

In 2002, the LTSN Generic Centre funded the preparation of a review to draw out lessons learned from a number of national learning technology projects. The projects selected were funded by UK higher education funding programmes over 1998-2002 and were predominantly TLTP phase 3 projects, but also included three ScotCIT projects and one FDTL phase 3 project that incorporated a significant ICT element. These projects had in common the aim to develop and implement innovative teaching and learning approaches supported by ICT and embed these within institutional practices. There were some distinctions in project aims between innovating and embedding. Some projects explicitly set out to develop or innovate and embedding was often a second order objective), whilst others were developed with the express purpose of embedding existing materials or tools in new contexts and were funded accordingly. The study applied a combination of qualitative evaluation methods: document review, semi-structured interviews and focus group techniques. An analysis of the outcomes and experiences has identified specific factors that these projects felt had influenced successful embedding of new practices within institutions and beyond.

Project Experiences of Institutional Embedding

Results from the study, particularly the semi-structured interviews, uncovered a host of common experiences in relation to institutional embedding. Most revealing, and in fact typical of what the projects themselves experienced in institutional embedding, is the comment:

We didn't meet the main formal outcome actually, so we couldn't tick that box but the real outcomes, the real impact, was much greater than what was originally intended.

If we therefore look beyond the project aims, the impact of the projects on their institution is in some cases quite considerable. The following discussion addresses some of the experiences that projects have identified as significant in a review of their own activities.

Exploring the potential of ICT
As far as the introduction of ICT is concerned, the projects performed an invaluable role in exploring its potential in learning and teaching. One project member remarked:

... if we hadn't gone through that experience we wouldn't have known what we needed or the sorts of questions we needed answered.

Decision-making processes, such as the choice of new technology or new learning situations, benefit from pilots and trials. The projects enabled their host institutions to conduct these pilots with external support. As one project evaluation report states:

During the interviews, statements such as "lots of opportunities to practice" have been repeated throughout and the process has been described as a supportive "playground" in which to try new things.

The projects were, therefore, an opportunity to gain knowledge about an area before committing institutional resources into ICT innovation.

Many of difficulties that project staff from inter-university consortia had in exploring learning technologies centred on working in institutions with an unfamiliar culture to their own. Furthermore, projects did not have a mandate to work in other institutions and were viewed suspiciously by local academics, information systems departments and staff developers alike. In their own institutions, projects inherited the difficulties central units themselves have in encouraging teaching development at both departmental and policy level. A strong tradition of initiatives and support from the centre greatly assisted many projects to work effectively across partner institutions.

Working within a supportive culture
Thinking about the culture in which embedding occurs, raises the issue of learning technology itself being a multidisciplinary field that draws on support from across an institution. The most effective embedding occurs usually where the institution and its departments have a supportive culture – where learning technology users do not feel isolated; where the relationship between the centre and the local is strong but flexible; and communications are good. There needs to be a balance between learning technology development and its support in practice to ensure that rapid developments in innovative areas can be co-ordinated by the institution. Two broad

models of support for learning and teaching innovation emerge from the project evaluation reports: a centralised development team that serves local needs; or local-ised developments which draw upon central and external support. What is clear is that support services and projects thus require 'two-way communication' to close the loops between central information strategic missions and local implementations.

There seem to have been assumptions made in projects about the role of support services and an almost tacit expectation that existing services within the institution will be able to extend their range to include new project requirements. In a couple of instances this was recognised and the original project teams were extended formally to include the input from library, computing service or staff developers. For exam-ple, the EFFECTS project noted that:

> at all five participating institutions, the support of staff involved … had been dispersed among a range of units including staff development, educational de-velopment, computing services and specialist learning technology units. (Smith & Oliver, 2000).

Integrating project staff expertise

Project involvement in itself brings benefit to the institution if the members of pro-ject staff are able to remain and consolidate their knowledge and experience. One project member noted that:

> There is now such a concentration of people [in the institution] who are now working in this area … that concentration of expertise is quite important.

The complexity of the learning technologist role has been documented elsewhere (Beetham, Jones & Gornall, 2001; Oliver, 2002). What is important to note here is the extent to which project reports highlighted the non-technical skills that learning technology project staff have to use in their work, including curriculum develop-ment, negotiation, advocacy, research, evaluation, dissemination, project and team management, resource planning and trouble-shooting. However, it has been long lamented that project staff are recruited on short-term contracts that expire with the lifetime of the project and that this generic expertise is often lost. Other factors in this regard relate to the nature of the project work being seen as separate from 'core business' and therefore an unsustainable 'development' activity.

Influence on institutional strategic thinking

While projects were recognised to have been helpful in informing institutional strategies for learning and teaching or for 'e-learning', there was not always a simple connection between a project and institutional thinking. It might be noted that from the outset, a number of projects set out to work with the grain of existing cultures and to drive existing institutional strategies and needs. For example, the TELRI pro-ject worked with research-led institutions to embed a research-based approach to learning supported through ICT (Blackmore, Roach & Dempster, 2002); the ELEN and ELICIT projects developed the use of integrated web based learning systems to deliver to key areas of institutional need; and EFFECTS worked with institutional professional development programmes to embed a reflective action learning approach to developing the use of learning technologies in teaching (Beetham & Bailey, 2002).

In relation to institutional missions strategies and policies, the influence of the project is sometimes hard to disentangle from general trends and movements in the institution. However, in a couple of the projects examined, there was felt to be some top-level impact and a sense that the project itself "got things moving and it supported change". A number of projects noted that some lecturers had their perspective transformed through participation. These individuals were labelled as innovators in learning technology, recruited onto relevant committees and went on to influence policy and thus embed the ethos of the project within the institution. The cascading of project knowledge and approaches from one colleague to another, one committee to another, within an organisation is a highly effective, longer-term change management strategy. Even though some projects labelled the strategy or policy impact as "coincidental", and most of those interviewed could not clearly isolate where project outcomes had been mainstreamed into the institution, there was documentary and focus group evidence in a number of projects that this had in fact happened. In one instance impact had been negligible and it was recognised that the project was "ahead of its time". Most of the other projects seemed to point to significant impact either at the level of the individual (lecturer) or at the institutional level (i.e. at institutions other than the host institution of the project).

Exchanging experiences across institutions and projects
The experiences gained through sharing practice and ideas with colleagues in other institutions was one of the most highly rated outcomes for the project teams to bring back into their own institutions. Project staff remarked that:

> …There's a whole new level of strategic thinking, which has come out of a lot of the work in [the project]. And not only from what we did within the University but also from looking at what others have done or are doing etc. I really can't say that all these ideas are our own, a number we picked up from other places.

Support for the projects in terms of the external funding or support agencies was generally commended. The main disjunctive area was felt to be inter-project collaborations. There was dissatisfaction expressed over the co-ordination of project outcomes and concern over longer-term survival of the knowledge gained. Some measures have been taken to address the issues by maintaining a central database of project outcomes, although it was recognised that this does not address the loss of impetus in embedding new practices after funding has ceased.

Lessons Learned From Successes in Projects

Given that most of the projects reviewed in the study (Dempster & Deepwell, 2002a) were completed over a year prior to the study, we have been able to extract a number of factors influencing the successful embedding of new practices within participating institutions and beyond. Often the criteria against which projects are deemed successful within the project community itself relates to process rather than outcome. An example of this is the high value project teams place on good collaboration with other institutions for problem-solving and sharing experiences. Whether or not the outcomes of the project are achieved, it is a positive experience if the collaboration has been good along the way. This in turn is a motivating force for continuation and embedding of project approaches into everyday practice:

> It's had a big impact for us as an institution, but also had a big impact on individuals. There has been a very vibrant, ongoing network, a kind of spiralling of people, building upon each others' knowledge and experience and what they've got from the project.

The link between 'buy-in' or ownership and successful adoption has been referred to frequently in the literature and was implicit in much of what projects experienced. Where projects are deemed to have "strongly influenced" strategic developments relating to the use of learning technology within their own institution, some common factors emerged from the research:
- *Timeliness* – in line with national and local developments, a number of inter-related institutional issues are being grappled with simultaneously and the timing for change can be quite critical;
- *Personal investment* – commitment beyond the letter of the contract, internalising project aims and investing in redefining the outcomes to ensure fitness for context;
- *Good collaborations* – networking, synergies, supportive culture across project team, across participating institutions and beyond;
- *Champions in policy positions* – making essential infrastructure changes and supplying further resources, taking on project/posts after external funding ceases;
- *Staff development* – integrating methods and materials into educational development for longer term gain;
- *Ability to adapt to local, emerging demands* – flexibility in interpretation of project outcomes, changing with the times.

Where these factors were in place, there was recognition that the project had embedded itself to a degree within the institution beyond the lifetime of the funding arrangements.

An Impact Model of Project Development

Based on the findings from projects on factors associated with successful institutional embedding, a number of generic frameworks emerge for thinking about project development (adapted from Dempster & Deepwell, 2000a):

- *Operational framework*- providing a means to consider the contexts in which projects mostly operate, assisting in identifying stakeholders and considering appropriate strategies for engagement and collaboration;
- *Planning framework*- aiming to assist projects in planning activities through their life cycle for implementation, collaboration, partnership development, evaluation and dissemination;
- *Organisational framework*- offering suggestions as to how these factors might translate into practical ways of optimising sustained impact of project approaches and "embedding" in teaching and learning practices.

In the context of the study (Dempster & Deepwell, 2002b), "projects", as referred here, are seen to be teaching and learning development projects, of an educational development nature, rather than research or product development projects, for which slightly different models might apply.

An operational framework
Associated with the issue of impact, distinctions in the operational contexts for consortium-based projects emerged from the project experiences of successful embedding:

- Stakeholder context: planning for impact- identifying stakeholders;
- Institutional context: in-reach activities: –informing local knowledge;
- Wider context: out-reach activities- making things happen elsewhere.

Externally funded projects serve many purposes within the academic community and are urged by the funding bodies to identify their stakeholder community. The stakeholders are well documented and range from the individuals and their communities of practice (e.g. learners, teachers, IT/library staff, educational and staff developers), the institutions and their senior managers, the development programmes and their reporting lines, national support agencies, and the funding bodies themselves. Figure 1 represents an operational framework for stakeholders of nationally funded projects and programmes.

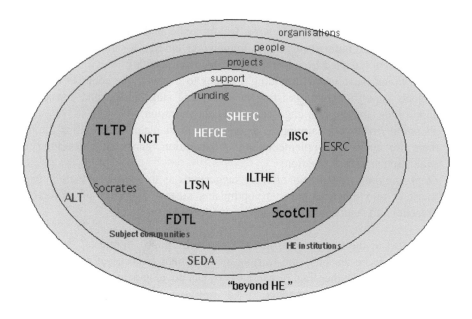

Figure 1. The project stakeholder context.

The interactions between institutional stakeholders within the project and intensity of the collaboration may vary (see Figure 2) and these might be worth considerng separately in terms of activities through the project lifecycle. Several different organisational models for consortia are apparent and this could usefully inform project organisation and priorities for activities. Collaborative approaches exist where various partners undertake different but complementary tasks. Parallel working describes a situation where various partners work independently but in parallel within a shared framework. A centralised approach would exist where the lead partner has major responsibility for development and assigns support activities to partners. Obviously, there are variations of these broad models and the approaches may also alter through the life cycle of a project.

The wider operational context might represent a combination of the stakeholder and institutional contexts. In representing this, a continuum emerges that maps out project organisation against specific stakeholders in terms of project activities (as shown in Figure 3).

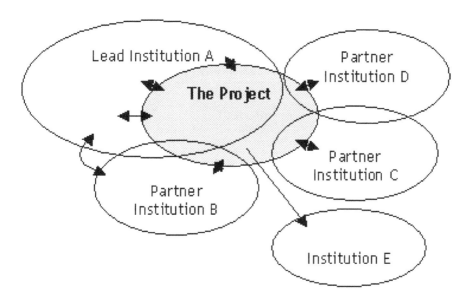

Figure 2. Consortium partnerships as institutional contexts.

A planning framework
Planning of project activities tends to focus on four strategies:
- Development/implementation;
- Evaluation;
- Dissemination;
- Exit/continuation.

The main development work might include production of educational concepts and tools, but predominantly the overarching aims of a national programme rely on implementation activities, in which particular approaches are tested out in specific teaching and learning situations. Staff development is often, but not always, secondary to implementation within courses, despite being an important part of the embedding process.

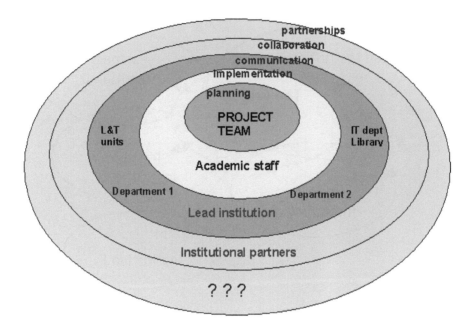

Figure 3. The wider operational context.

Projects that have been successful have looked at these elements in a holistic and non-linear manner. Planned activities do not necessarily start at the centre and proceed systematically to the outer layers. Planning is aimed at engagement across operational, institutional and wider contexts simultaneously from the outset. Project activities aim at synergy between these strategies and are planned for throughout the project lifecycle. This maximises the likelihood of sustained impact of the successful new approaches developed by the project.

An organisational framework
An early evaluation of TLTP projects by the Tavistock Institute (Sommerlad et al, 1999) focused on four key "implementation" strategies that support the activities identified by projects as successful in terms of institutional change. These are:
- Negotiating entry and pitching in at the right level;
- Securing institutional support and getting the right stakeholders on side;
- Mobilising and engaging teaching staff and other key actors;
- Diffusing technology based teaching and learning innovations.

Building on guidance for dissemination from the TQEF National Co-ordination Team (TQEF NCT project briefing no. 2), we suggest planning activities according to the following four objectives:

1. Awareness raising;
2. Increasing understanding;
3. Increasing uptake;
4. Embedding.

The view is that activities that lead to the permeation of new ideas at the ground level that might lead to implementation and sustained changes in teaching approaches require some kind of scaffolding. This is a useful form of impact hierarchy, assuming that lecturers will move forward through the four stages from thinking to doing. The first three facilitate some degree of impact on practice, but do not necessarily lead to sustained changes in practice. Certainly, projects should be organising the activities with all four goals in mind, but there is not a guarantee that each will lead to the next level in terms of impact.

By systematising each activity across each layer of the project's wider operational context (Figure 3), projects might aim to address all four objectives through a strategic approach for linking current activities and outcomes with forward-looking ones. Figure 4 shows how a single planning activity at the project team level might be transferred concurrently throughout the local, institutional and wider contexts.

Guidelines for strategic project development
The aspects of project organisation, planning and management described in the previous sections are typical in projects that have:
•   Had a significant impact;
•   Produced sustained changes in practice;
•   Embedded their approaches widely.

The review of educational development projects identified a set of "success factors", while the impact model for project development suggests a generic framework. Taken together, these offer, in effect, guidelines for strategic project development.

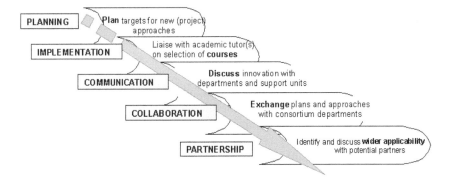

Figure 4. Strategic project development: transfer of project activities from the local to the wider context.

Planning - direction and commitment:

- Develop project aims and activities in line with local strategies and objectives where possible;
- Involve schools and departments directly in the creation and delivery of the project;
- Recruit enthusiastic individuals or second those known to be committed and who understand the nature of project work;
- Build project and partnerships on the timely needs of the field, the sector or the institutions.

Implementing - engagement and implementation:

- Take a flexible, modular approach to sub-project developments where possible, so that if one has to be dropped, others can take their place – the Pareto Principle;
- Avoid time-wasting on non-committed sub-projects by having back-up;
- Provide written guidance on new techniques or software and aim to empower academics, departments and the project as much as possible in the use of technology;
- Ensure that those not directly involved in the project (at least from the outset) are involved early on; market the project well from the start – it really pays off.

Communication and collaboration:

- Develop a "community of practice" or user group/special interest group around the project activities and development;
- Look early to the wider applicability and plan for it in development, evaluation and dissemination activities – perhaps make use of an experts group or evaluation focus group throughout the project. This also serves to network those involved and bring in a wider range of interests and experiences;
- Ensure regular liaison between sites, and that communication is inclusive of all project staff and local support;
- Involve all sites in evaluation activities, which also serves to enrich the data collected.

Partnerships, roles and relationships:

- Ensure a clear delegation of activities and responsibilities between project directors –, project manager, and educational developers – as "owners" of the project, academic leaders, operational managers, day-to-day co-ordinators, technical developers, etc.;
- Aim to involve other support departments closely from the outset of the project (even in the bidding process), particularly IT departments for ICT-related projects;
- Ensure continuation of knowledge and skills by professional development, joint activities and good documentation.

Conclusions

The focus of our study of national learning technology projects was not so much whether the projects successfully met their intended outcomes, but rather how they successfully effected changes within their home institution or within collaborating institutions. The aim was to identify common features of these projects that may have facilitated these changes. Two broad dimensions of the success factors could be discerned: in-reach and out-reach.

In-reach is where the project has embedded its ideas or approaches or tools within the home institution and informed academic practice. Ideas and concepts have been cited in this regard, materials and tools seem to have been less widely taken up.

Out-reach is where the project has influenced change elsewhere. Some of the projects felt that they may have actually "failed on home territory" and yet they were certain that their influence had been significant in other institutions. Many project directors remarked that projects had been important in assisting or even opening debate about teaching and learning practice, or aspects of it, and the role of technology in these processes. The influence of the project nevertheless frequently became entangled with general strategic development within institutions.

There are a number of factors common to most projects, which appear to be key drivers in determining embedding within institutions. These factors fall across three broad motivations for involvement in projects:
- external (response to national initiatives or funding streams);
- internal (response to institutional or department strategy or targets);
- self-directed (personal interest of project innovators).

Where these motivations converged within a project, there seemed to be the greatest likelihood of successful embedding of both the ideas and the approaches within institutions. In these cases, the outcomes have the potential to be long lasting, to be an integral part of staff development programmes and to inform the strategic direction of the institutions targeted.

Experience across the national projects reviewed in this study suggests that projects that are working with the grain of the institutional culture are more likely to embed change. It is perhaps a little ambitious to expect that significant evidence of embedding be produced within the period of a project's funding. Most changes in practice take place over time and in the review of projects, it is apparent that this can vary quite substantially between projects and programmes. Post-project evaluation might, however, usefully address:
- Approaches that led to good impact in the institutions;
- The kinds of local and external activities that were effective;
- How the project is moving from innovating to embedding.

Final Thoughts

Given the outcomes of the review activities in this study, it is interesting to reflect on the ways in which the funding of national projects is an effective approach to developing and embedding new practices in teaching and learning in higher education. Many of the projects reviewed were funded with the express purpose of embedding rather than with the aim of innovation and development. Nevertheless, the distinction and distance between innovation and embedding is vast in regard to the processes of educational development and institutional strategic development. The capacity of an institution to ready itself to respond to new approaches that technology or other innovations can offer is also paramount.

National projects are not always resourced, equipped or best positioned to handle the task of organisational change as they generally have insufficient linkages with strategic planning and institutional mechanisms. At the onset of TLTP phase 3, the Tavistock review (Sommerlad et al, 1999) had already highlighted an issue that persists today, that:

> this does raise questions about a broad programmatic strategy of implementing generic products and services across diverse institutional contexts, unless there is very real scope for customisation and local embedding.

One might argue that the vital issue is not whether learning and teaching development in supported via national or local funding. Rather, the role and integration of local support networks is crucial to successful embedding of new modes of practice beyond the individual. A number of models for staff development in e-learning have been recognised as effective (Oliver & Dempster, 2003) and most of these rely heavily on central or departmental support, for example in terms of the technology infrastructure, funding and incentives and staff training and guidance. Those facilitating institutional embedding must explore, nurture or initiate forms of dialogue that can support effective sharing of practice. A first step at the local level might be to find out what lecturers and departments feel is important (in their discipline) in developing their teaching and enhancing their students learning experiences. Such evaluation may suggest areas already identified as key issues or problems for which generic solutions might be applied.

References

Beetham, H., & Bailey, P. (2002). Professional development for organisational change. In R. Macdonald & J. Wisdom (Eds.), *Academic and Educational Development: Research, Evaluation and Changing Practice in Higher Education* (pp. 164-176). London: Kogan Page.

Beetham, H., Jones, S., & Gornall, L. (2001). Career Development of Learning Technology Staff: Scoping Study Final Report. [On-line]. Available: http://sh.plym.ac.uk/eds/effects/jcalt-project/final_report_v8.doc

Blackmore, P., Roach, M., & Dempster, J. A. (2002). The use of ICT in education for research and development. In S. Fallows & R. Bhanot (Eds.), *Educational Development through Information and Communications Technology* (pp. 133-140). London: Kogan Page.

Dempster, J. A., & Blackmore, P. (2002). Developing research-based learning using ICT in HE curricula: the role of research and evaluation. In R. Macdonald & J. Wisdom (Eds.), *Academic and Educational Development: Research, Evaluation and Changing Practice in Higher Education* (pp 129-139). London: Kogan Page.

Dempster, J.A., & Deepwell, F. (2002a). Enhancing the successful embedding of project outcomes: A Generic model. [On-line] Available: http://www.telri.ac.uk/Transfer/ltsngc/Outcomes/Generic_model.pdf.

Dempster, J.A., & Deepwell, F. (2002b). A review of successful project approaches to embedding educational technology innovation into institutional teaching and learning practices in higher education: A Study funded by the LTSN Generic Centre. {On-line] Available: http://www.telri.ac.uk/Transfer/ltsngc/Outcomes/Review_report.pdf.

Oliver, M. (2002). What learning technologists do. *Innovations in Education and Training International*, 39, 4, 1-8.

Oliver, M., & Dempster, J.A. (2003). Embedding e-learning practices. In R. Blackwell, & P. Blackmore (Eds.), *Towards Strategic Staff Development 9pp. 242-257).* Buckingham: SRHE/Open University Press (in press).

Smith, J., & Oliver, M. (2000). Academic development: A framework for embedding learning technology. *International Journal of Academic Development*, 5, 2,129-137.

Sommerlad, E., Pettigew, M., Ramsden, C., & Stern, E. (1999). *Synthesis of TLTP Annual Reports.1999.* Unpublished report, Tavistock Institute.

Taylor, P.G. (1998). Institutional Change in Uncertain Times: lone ranging is not enough. *Studies in Higher Education*, 23, 3,269-79.

TQEC (Teaching Quality Enhancement Committee) Report (2003). Final Report of the TQEC on the Future Needs and Support for Quality Enhancement of Learning and Teaching in Higher Education. [On-line]. Available: http://www.hefce.ac.uk/Learning/TQEC/

TQEF NCT (National Co-ordination Team). Project Briefing No. 2: Dissemination. [On-line]. Available: http://www.ncteam.ac.uk/resources/project_briefings/briefings/brief02.pdf

Projects Reviewed

TELRI (*TLTP3/no.92*) http://www.telri.ac.uk, Technology Enhanced Learning in Research-led Institutions

ASTER (*TLTP3/no.94*) http://cti-pys.york.ac.uk/aster/, Assisting Small-group Teaching through Electronic Resources

SoURCE (*TLTP3/no.79*) http://www.source.ac.uk/, Software Use, Re-Use and Customisation in Education

EFFECTS   (*TLTP3/no.89*) http://sh.plym.ac.uk/eds/effects/, Effective Framework for Embedding C&IT using Targeted Support

TALENT (*TLTP3/no.82*) http://www.le.ac.uk/TALENT/, Teaching And LEarning with Network Technologies

ELEN   (*TLTP3/no.84*) http:// www.lincoln.ac.uk/elen/ , Extended Learning Environment Network

ANNIE   (*FDTL3/no.60*) http://www.ukc.ac.uk/sdfva/ANNIE/, Accessing and Networking with National and International Expertise

SESDL (*ScotCIT*) http://www.sesdl.scotcit.ac.uk/, Scottish Electronic Staff Development Library

ELICIT (*ScotCIT*) http://www.elicit.scotcit.ac.uk, Enabling Large-scale Institutional implementation of C&IT

NetCulture *(ScotCI)* http://netculture.scotcit.ac.uk/, Staff Development Network

Acknowledgements

The authors are very grateful for contributions to this study by Peter Jackson who undertook the telephone interviews and by Mark Childs who extracted aspects of the guidelines to project organisation, planning and management.

5

Embedding Learning Technologies into Institutional Practices: A Further Education Perspective

Joe Wilson

This chapter will attempt to set out the position that learning technology in further education (FE) was at ten years ago, review the changes and drivers and finally look to the future and the ways that the further and higher education sectors can learn from each other. The scale of this task necessitates taking a broad overview of a myriad of initiatives and reports from a range of agencies operating at local, regional and national levels. Inevitably this means all projects and initiatives cannot get equal coverage. In compiling this piece I consulted with colleagues in all the national regions and I hope it captures the essence of an exciting and vibrant period in the development of learning technology in the FE sector.

In setting the context it should be noted that it is hard to generalise about further education colleges in the UK. Ten years ago most UK colleges became incorporated bodies. Cut off from their traditional local authority managers FE colleges were forced to radically restructure and adopt corporate practices. The ethos in the FE sector became competitive and expansionist, conditioned by the resources available from the funding bodies and those available from other sources.

Recently parliamentary devolution has created more subtle distinctions in how FE is delivered and funded in the UK. The agencies, policy frameworks and initiatives at a national level in FE are different and distinct in the different national regions of the UK. If one generalisation can be made, it is that across the UK, FE colleges at their best are dynamic agents of change who have had to develop flexible curriculum provision to meet the needs of learners and the requirements of funding bodies.

Setting the Scene

Ten years ago I was working as a section leader in a medium sized FE college on the south side of Glasgow. The college had installed a network system to support the administrative functions and e-mail to allow managers to communicate with finance and other services. Teaching staff had limited access to PCs through a staff resource base and Internet access for the college was limited to some dial up accounts.

Staff development resulted in a few enthusiastic individuals who completed Internet modules word-processing, PowerPoint and perhaps some desktop publishing units and went on to monopolise the machines available in the staff resource base. Some external training was available from now extinct organisations like MEDC (the Microelectronics Educational Development Centre) based at Paisley University, which offered some specialist training in a range of computer applications.

Resources tended to be CD based or videodisks, for example a newspaper archive on CD or 'Europe in the Round' [1]. Access to these resources, however were hampered by lack of CD ROM servers and general lack of machines in library resource areas. Learning resource development focused on developing distance learning or flexible learning materials usually in partnership with COLEG (Scottish Open Learning Exchange Group) [2]. To facilitate resource development, some colleges developed specialist publication units, with a particular focus on desktop publishing. This pattern was similar around the UK with NCET (National Centre for Educational Technology now BECTA) and FENC [3] (Further Education National Consortium) supplying the training and developing flexible learning materials in England and Wales.

With the exception of those staff who worked for colleges who were members of NILTA (National Information and Learning Technologies Association) ten years ago a learning technologist would have been regarded, as the technician who sporadically changed the bulbs of the department's limited overhead projectors or the member of teaching staff who really knew how to work the college photocopier.

Policy Drivers

In the mid to late 90's a range of influential reports began to appear, all of which featured further education and suggested to a greater or lesser extent that learning technology could have a significant impact both on delivery and how institutions should manage learning. At a UK level these reports included:
- Dearing: A review of qualifications for 16-19 year olds (Dearing, 1996);
- Beaumont: A review of National Vocational Qualifications (Beaumont, 1996);
- Tomlinson: Principles of inclusive learning and recommendations (Tomlinson, 1996);

- Higginson: The use of technology to support learning in colleges (Higginson, 1996);
- Kennedy: Widening participation (Kennedy, 1997);
- Fryer: Lifelong learning (Fryer, 1997);
- Dearing: Higher education in the learning society (NCIHE, 1997);
- Connecting the Learning Society: The National Grid for Learning (DfEE, 1997);
- Opportunity Scotland: Lifelong learning (The Scottish Office, 1998).

With the benefit of hindsight it can be seen that the Higginson Report (1996) was perhaps the most influential of these in respect of learning technology. The five main recommendations of this report were:

- Establishment of a National Staff Development Programme to develop information, communication and technology (ICT) skills in sector and to banish scepticism;
- The creation of National Learning and Technology Centres linked to curriculum development;
- Funding for a range of demonstration projects (e.g. management information systems, student tracking, IT assessment, online tutorials , video conferencing etc.);
- Establishment of a managed communication network;
- Establishment of a programme of research and development.
- Early funding initiatives

Despite the array of policy documents that came out in the mid to late 90's national funding did not immediately appear to support this vision. Colleges across the UK were therefore seeking funding from other sources. Initially, colleges discovered that they could access European Funding through the Adapt Project [4] and access support through individual projects like Ariadne [5]. ADAPT was a Human Resource Community Initiative funded through the European Social Fund with a budget for the period 1995-2000 of €333.5 million in the UK. Introduced in 1995, ADAPT supported projects, which aimed to help workforce development. In the UK, there were three bidding calls or rounds of the ADAPT Initiative. Funds from projects like Ariadne and ADAPT allowed colleges to explore the potential of e-learning through collaborative activities. Colleges explored e-learning in the hope it would allow them to attract and retain learners in an increasingly competitive environment.

Early Regional and National Developments

In Scotland regional partnerships formed, principally GTN, Wessnet and Sesnet. Initially these partnerships were formed to provide cost effective regional network infrastructures but soon encouraged staff to develop online learning materials. European Union funding for these partnerships was matched by local development agencies and other local and national partners.

The GTN (Glasgow Telecolleges Network) was a partnership featuring the ten FE colleges in the city, supported by the four local universities, the Scottish Office,

The Scottish Council for Educational Technology, the Glasgow City Council and the Glasgow Development Agency. By 1998 this broadband network supported 85,000 students and offered online learning, video conferencing and access to data service.

Colleges also began to explore virtual learning environments and mechanisms for creating and managing online content. These projects explored the use of Lotus Learning Space (and other commercial virtual learning environments) as well as the use of 'Pioneer' a learning environment developed and freely available from SCET (Scottish Council for Educational Technology now Learning and Teaching Scotland).

These regional developments led to some national responses. For example, the Scottish Learning Network (SLN) was an ambitious project to support the fledgling Scottish University for Industry [6]. It included in its targets the provision of online learning opportunities for 5,000 learners and training for 1,000 tutors to support on-line learning. Online course materials were produced and many FE staff engaged in SLN training (although few completed the programme).

The Scottish Further Education Unit developed the VLC (Virtual Learning Centre) [7]. The aim of the VLC was to support staff involved in the delivery of vocational education and training in Scotland by providing an information gateway to assist staff in identifying electronic resources suitable for use by students, and online staff development courses for lecturers and trainers. The VLC project provided a staff development opportunity for staff from across Scotland and led to the creation of a shared electronic resource that mapped learning materials onto the Scottish FE curriculum.

In England and Wales the QUILT project 1997-199 (Quality in Information and Learning Technology involving both FEDA (now the Learning and Skill Development Agency) and BECTA (British Educational Communications and Technology Agency) allowed colleges to access project funds [8]. The QUILT programme encouraged colleges to engage in learning technology projects, organised one day events, offered executive briefings, organised a problem-solving network and identified some colleges as exemplars of good practice. Colleges were invited to bid for project funds under a number of different themes: development of Information and Learning Technology (ILT) skills and integrating ILT into the curriculum, using ILT for flexible learning, developing staff to support on-line learning, creating an Intranet or Extranet, developing electronic learning resources and using ILT to enrich class teaching.

Birth of the Learning Technologist

Colleges relied on home-grown expertise or advice from organisations like the Scottish Council for Educational Technology (SCET), NILTA, BECTA, Further Education Development Agency (FEDA) and the Scottish Further Education Unit (SFEU) to support developments in the use ICT. Gradually learning technologists began to emerge from the sector and find a focus for their work. For example, in Scotland The ALTAC (Applying Learning Technologies Across the Curriculum) Conference became a meeting place for the emergent learning technologists in Scotland in the mid nineties and was supported by the SCET. The main focus for UK wide gatherings became the NILTA or BECTA conferences. At these gatherings we often learned from our failures as much as our successes.

Staff development opportunities arose through becoming involved in EU or nationally funded projects. For example, some staff discovered the LETTOL (Learning to Teach On Line) programme developed by Sheffield College through ADAPT funding as a route to a more formal qualification [9].

Later Funding Initiatives

By the late nineties learning technology had made an impact on UK further education but only in those colleges that had been able to engage in local projects. A large investment was still required to meet the recommendations of the Higginson Report. A significant minority of colleges still did not have reliable access to the Internet, many more lacked a local network and most staff lacked basic IT skills. Learning technology was still a 'minority sport'.

In 1999, colleges anxiously eyed the development of the Scottish University for Industry and the University for Industry (UFI) in England. These agencies were funded to respond to the needs of learners and employers in meeting their skills needs through the exploitation of information communication technology which resulted in substantial funding finally being injected into the FE sector in Scotland, England, Northern Ireland and Wales. In England £74 million was injected into the sector in the period 1999-2002.

In Scotland the new Scottish Executive made £29 million available to the sector over the period 1999-2002 for the development of the further education component of the National Grid for Learning. Comparable figures were made available to colleges in Wales and Northern Ireland. This period of large and sustained investment led to a boom in developments at college level across the UK and a raft of further national and regional initiatives.

Linking Strategy to Policy

In England the Further Education Funding Council published two key documents: 'Networking Lifelong Learning Making it Happen' followed by an ILT Implementation Plan (Further Education Funding Council 199a,b). These strategic planning documents set out a range of priorities, much of which built on the QUILT experience. Other funding bodies were also putting mechanisms in place to respond to policy drivers (See Table 1).

In Scotland, the Scottish Further Education Funding Council (SFEFC) managed the disbursement of funds at a strategic level. The Council's priorities developed through consultation with the sector and were presented to the sector as SFEFC's ICT Strategy [10]. The Council's priorities were the development of a national infrastructure, supporting the development of local infrastructure, staff development, materials development and supporting the development of managed learning environments. These priorities were broadly similar to those of England, Wales and Northern Ireland.

To support local infrastructure developments colleges were required or encouraged to produce strategic development plans mapping out their local development plans. In some areas of the UK, funding was dependent on the submission of an approved plan. Across the UK, colleges received funding to develop their networks, increase the machines available on these networks and fund ICT staff development activities. Targets were set by most funding authorities based on ratios of full time staff and students to the number of networked computers available. Alongside the hard targets, softer targets were set, encouraging college staff to develop basic IT skills. In Scotland these were expressed as encouraging staff to gain The European Computer Driving Licence (ECDL) or equivalent qualifications as a baseline. In Scotland and England, progress was monitored by BECTA surveys charting the development of ICT provision in the sector. The surveys showed colleges using the ring-fenced funds to improve their networks, invest in virtual learning environments and increase the ratio of networked PCs. The surveys do not give much detail of the impact of the investment on the curriculum or on learners [11].

Later Regional and National Developments

In addition to the provision of wide area network connections, training and resources to enable colleges to strengthen their local area networks, a number of significant research, support and development activities were initiated. For example, membership of JISC allowed colleges to partner with HE institutions in a range of calls and many have benefited through access to JISC project funds and services. It is impossible to list them all here, but notably X4L has been successful in allowing a range of FE and HE Institutions to work together [12].

In Scotland, SFEFC contracted the ICT Staff Development and Advisory Service (SDAS) and the Content Development Advisory Service (CDAS). These services supported the Scottish colleges and allowed a fruitful partnership to develop between the existing agencies who supplied the support services and the Scottish JISC Regional Support Centres. Agency collaboration produced a successful MLE/VLE conference, a large programme of technical training to support the staff managing local college networks, the first attempt at a sector wide Training Needs Analysis led by the JISC Regional Support Centre North and East Scotland, and the Scotfeict site hosting a database of national training opportunities to provide cost effective training to college staff and a range of other activities [13]. CDAS contributed to many of these activities, mapped the NLN content on to the Scottish Curriculum and provided liaison with between NLN content developments.

Table 1: Policies and mechanisms put in place by funding bodies across the UK

Priorities	Example Responses by Funding bodies
National Infra-structure	Buying service from Joint Information Systems Committee (JISC) and UKERNA (United Kingdom Education and Research Networking Association) supplying 2 mega bit link for colleges. In some regions this was soon upgraded. Encouraging existing FE support agencies to work with JISC services
Local Infrastructure	Giving colleges ring fenced funding based on formula funding allocation to invest in local network and to increase the ratio of networked PCs for staff and students. Making training made available for network and technical support staff. Encouraging local planning by requesting ICT development plans.
Staff Development	Giving colleges ring fenced funding based on formula funding allocation to invest in staff ICT/ILT staff development. Creating new training organisations in regions (JISC established 13 Regional Support Centres). Developing 'train the trainers' initiatives (e.g. ILT Curriculum Champions programme, FERL Practitioners Pack). Creating or further developing supporting web sites and online courses (FERL, NLN, SCOTFEICT,SFEU). Encouraging a range of suppliers by setting up national training opportunity databases. Using 'sweeteners' (e.g. computers for teachers scheme). Encouraging local planning, by requesting staff development as an element of ICT development plans. Promoting the development of some bench-marks or standards for ILT/ICT staff development.
Materials Development	Central purchasing of materials: e.g. NLN content - private sector developers. Tailoring existing JISC content. Encouraging some local developments and reuse through project funding from JISC and other sources.
Virtual or managed learning environments	Encouraging colleges to purchase or explore through ring fenced funds. Arranging national events to showcase good practice and available products. Tapping into JISC and HE expertise. Encouraging VLE or MLE by asking for it to be part of development plan.
Encourage innovation research and development	Inviting colleges to bid for funds either on basis of specific calls or on sets of national / regional priorities. Encourage Award Schemes to reward best practice e.g. ASC Beacon Awards, BECT ICT in Practice Awards, NILTA Awards. Showcasing good use of technology at events such as BETT and SETT and through web sites such as FERL and Scotfeict. Encouraging HE and FE projects to develop in response to JISC project calls allow FE and HE project partnerships to develop.

In addition to this activity SFEFC funded a range of strategic projects in a range of areas spanning; online assessment, diagnostic testing of core skills, use of video-conferencing for guidance, web-based guidance tools, online enrolment, Intranet development and management of flexible learning centres.

In England, the National Learning Network (NLN) was established [14]. The NLN initiative is managed by BECTA and delivered by a strategic partnership of sector bodies including: Learning and Skills Development Agency (LSDA), JISC, NILTA and UKERNA. In summary, the activities in the NLN programme involve:

- Connecting all colleges to the JANET infrastructure by 2001 (Lead by UKERNA);
- Setting up the Regional Support Centres (JISC);
- Researching the specification of MLEs (JISC);
- Procuring and developing national learning materials (BECTA);
- Setting up an ILT subject mentors programme (BECTA);
- Setting up an ILT champions programme (BECTA);
- Developing materials standards (BECTA);
- Developing ILT standards (LSDA);
- Supporting innovative projects (LSDA);
- Evaluating the NLN (LSDA).
- Auditing and mapping the content and materials (LSDA);
- Setting up a technician training programme (FEFC/LSC);
- Investigating the feasibility of a central resources bank (NILTA);
- Measuring LAN standards (NILTA).

A glowing evaluation of the NLN was conducted in 2002 by the LSDA[15]. The evaluation showed the positive effect that the NLN investment has had on learners and colleges and highlighted some of the future challenges. The main challenge is that of matching the resources available to the growing demands from staff and students for access to learning technology.

In England, BECTA developed the FERL web site as a useful sector resource supporting the ILT Champions programme and the FERL practitioners' programme[16]. The FERL web site provides information for lecturers, trainers, managers, ILT Champions and support staff. It offers advice, guidance, and examples on the use of ILT in all aspects of further education. The site also provides resources for use in teaching and learning, with guidance on how these could be, and have been, implemented.

BECTA also managed the 'Computers for Teachers FE Scheme' on behalf of The Learning and Skills Council [17]. This scheme offered selected teachers in FE and 6th form colleges in England, a subsidy of 50% towards the cost of a personal computer (from an accredited supplier) up to a maximum of £500. The Learning and Skills Council also committed to paying any income tax and National Insurance due on these purchases.

A Growing Community of Learning Technologists

To support staff development across the UK, The Further Educational National Training Organisation (FENTO), the national lead body for the development, quality assurance and promotion of national standards for the FE sector produced the ILT Standards. These were a set of standards relating to those involved in teaching and supporting learning and management in FE using new and established technologies and were developed following widespread consultation with further education managers, staff developers and teachers, through a series of detailed trials and workshops.

The FENTO standards provide a framework within which to identify the competences and knowledge necessary to perform effectively with the help of new technology. They are designed to provide essential diagnostic and planning information in the ILT strategic planning process. The expectation being that staff who have the European ECDL, or equivalent, will now develop base line ILT skills. Existing qualifications have started to align themselves to these new standards (e.g. Lettol). In Scotland the SQA have developed a continuing professional development qualification mapped to the FENTO standards, which will be launched in June 2003. Increasingly ILT or ICT staff development will be part of the continuing professional development landscape for staff working within UK FE.

The push to improve skills and standards has contributed to the development of an established community of learning technologists in the FE sector. National gatherings are supported by a range of FE agencies and online communities with interests in a range of issues are emerging (using JISC mail services). The Association for Learning Technology (ALT) is now seen by many in FE as an important interface between learning technologists in FE and HE. For example, many JISC collaborative projects have grown out of meetings at ALT conferences or events. Notably in Scotland a number of successful X4L and 'Fair' bids have been pulled together either at or on trains to and from ALT events.

Future Challenges?

The past ten years have been a dizzying period of development for FE. This period of investment has allowed most colleges to produce an ICT development strategy, purchase or develop virtual learning environments and start to wrestle with the creation of a Managed Learning Environment (MLE). They can access appropriate training from a range of agencies, access content through NLN and other projects and map their staff skills base against FENTO or other standards.

A major consequence of this period of high profile and high investment is that the inspection bodies in UK are now developing inspection frameworks to evaluate the use of ICT in the classroom while the funding bodies are beginning to look for harder evidence of a return on this investment. The challenge for many FE institutions will therefore be to sustain and embed the advances made. There is a danger that some colleges have bolted on ILT developments rather than made them central to their mission. If colleges are to optimise the investments made in infrastructure, staff development and content development they have to be embedded in a whole organisational strategy.

Staff development needs to move beyond even the new FENTO ILT standards. Subject specialists will need to continually update their skills and share best teaching practice with one another. Online communities of practice need to be fostered and developed.

The immaturity and volatility of some learning technology mean that there is a lot of work involved in keeping up with available products. Much effort has been wasted through poor understanding of the technology and its application. FE is reliant on HE and JISC services with their greater resources to insure that colleges do not make expensive mistakes. Services like JISCInfonet, a service 'delivered by a partnership of FEIS and HEIS, has been developed to support the FE and HE sector in planning and developing cost effective MLEs[18]. Increasingly FE will become reliant on information from bodies like CETIS [19] and services like Techdis [20] to steer them through the standards and specifications that are evolving and hard to understand.

FE will look to HE for useful and practical research and case studies to support an understanding of the theory and pedagogy that underpins the use of the technology. FE will also look to HE, in the first instance, for an established and practiced methodology to rigorously evaluate e learning and for higher-level qualifications for staff. It is also likely that to sustain and develop network services, including the future provision of Virtual Learning Environments (VLEs), colleges will need to work much more collaboratively with one another and with HE partners to squeeze the best value out of the available technology. There are a host of areas where shared network services could lead to service improvements for learners and cost savings for institutions.

Some projects indicate ways in which the FE and HE sectors are already working together to provide cost effective network services, the Welsh Video Network is the largest single IP (H.323) videoconferencing network in Europe, and one of the most advanced in the world [21]. In Northern Ireland, The Northern Ireland Integrated Managed Learning Environment (NIIMLE) aims is to build a cross-institutional MLE for the Northern Ireland region that facilitates the mobility of the lifelong learner and supports collaboration between FE and HE institutions [22].

But it is not a one-way street, HE can learn from FE about employing a range of learning and teaching strategies and managing a more accessible curriculum. FE has a breadth of experience in developing and delivering a flexible curriculum of modular courses around national vocational competency frameworks and in dealing with large cohorts of part-time students. In some cases HE can learn from FE about the large-scale deployment and adoption of VLEs. Much good would come if FE and HE subject specialists, librarians and learning technologists were encouraged to collaborate and share best practice in more formalised ways.

The immediate future of ICT developments in the FE sector are mapped out in 'Success for All' (DFES 2002a), the 'Get on With It' Report (DFES 2002b), DELG Recommendations [23] and in Scotland SFEFC's strategic plan.[24] In Wales a major consultation exercise is currently underway to map out an 'e-learning strategy for Wales [25]. In some respects these plans map out more of the same. 'Success for All' looks to a more co-ordinated response in the continuing expansion of e-learning in the post 16 curriculum through the creation of better mechanisms to share online content. 'Get on With It' addresses the challenges and opportunities in taking online learning into the workplace. The DELG recommendations focus on how online learning can support local skills council needs. While SFEFC's plan looks at ways of getting best value from current investment.

Much of the current debate at national and institutional levels are around common themes; the mechanisms for sharing or reusing learning objects, the improved management of the learning process (really the administrative process) that can be gained through the development of a managed learning environment and the problems in getting staff to upgrade their skills to cope with all of this technology. But the debates and initiatives still echo the Higginson Report while technology and many learners have moved on.

A Personal Reflection

When I started using learning technology with my students ten years ago I did it because it was exciting and new for me and for the learners. Internet technology offered them access to a wealth of resources that my college would not have been able to provide in other ways. A course web site enabled learners to access my classroom materials when they wanted. A chat-room provided them with peer support and an area where they could talk on or off topic as they pleased. I saw technology as a mechanism to liberate learners, open minds, empower. Ten years ago a course web site, a chat-room offering threaded discussion and the facility to e-mail your tutor were really new things for many students.

Ten years on the deployment of a VLE and the development of a MLE formalises many of these developments and makes these support mechanisms a part of the institutional fabric. The large-scale adoption and deployment of systems like this make staff development and the management of learners easier but there is a danger that the systems take the creativity and fun out of both learning technology and learning. VLEs do usually make everyone's course look the same. This can be hugely gratifying for college quality managers and make the processing of returns to funding bodies easier, but it might not be that rewarding for students or tutors in the long run (I can imagine how thrilled I'd have been as a first year undergraduate to be told that the virtual learning environment would track all my learning interactions).

Over the years as I have demonstrated and enthused about the Internet, online courses, better presentation software, college intranets, ways to create online assessments, electronic whiteboards, streaming and digital video, and more recently wireless networks. I am always struck by the fact that in the main those that come along to the workshops want to use technology to improve how they currently teach or lecture but only a minority are interested in how technology should actually be changing educational practice and how we do things.

I groan at the thought of students faced with death by PowerPoint both in the lecture theatre and now in the VLE. They will also really 'enjoy' the endless automatically generated e-mails they will soon be sent asking why they have failed to submit their assessments on time.

It is interesting to see how much we are stuck on the formal transmission modes of learning and our institutionalised control mechanisms. We haven't really got our heads round how students are using the technology yet and really thought about how we could use technology to allow learners to do far more active learning.

Some of the more interesting work I have seen, has involved giving students access to the technology and pushing the boundaries a bit. For example, encouraging them to create their own multi-media Weblogs then using the whiteboards and presentation software to deliver to each other; the use of mind-mapping software with teaching staff and students to look at new ways both of structuring and interpreting course structures; the use of file sharing to peer assess essays and assignments and se of cheat sites on the internet as sources of essays to be graded and marked as part of the learning process. The pockets of really interesting work I encounter, challenge learners (as well as sometimes the local network security policies!).

Conclusions

The implementation and integration of ICT into FE has developed at a pace, driven by policy initiatives, massive funding opportunities and strategic planning. In the future, it would be interesting to see some policy initiatives that focused more on the learner than the technology and started to set out ways that learners could:

- design and define their own learning;
- design and maintain their own virtual learning space;
- develop virtual portfolios of their work;
- collaborate both within and across the institutional frameworks;
- influence their own and others learning
- have the right to summative assessment on demand.

Perhaps for all this to happen the lines between schools, FE, community education, HE and the world of work need to be much more blurred. In Scotland the focus is very much on life long learning and bridging the gaps between school, FE, HE and community education. Consultation has already begun on a merger of Scottish Further and Higher Education Funding Councils following the recent report by the Enterprise and Life Long Learning Committee [26]. In the rest of the UK, the debate is now about post 16 education and HE. The blurring of the lines may be about to happen.

The expectation is that learning technology will move from being the domain of the few to being embedded in good practice, part of entry-level qualifications and continuing professional development for teaching staff and part of the culture and fabric of learning in the 21st century. To really make it part of the fabric, learning technology needs to become the learner's technology. That 's the challenge for the next ten years.

References

Beaumont, G. (1996). *A review of 100 NVQS/SVQS (Beaumont Report)*. London: DfEE.

Dearing, R. (1996). *A Review of Qualifications for 16-19 year olds*. SCAA Publications.

DFES (2002a) *Success for All. Reforming Further Education and Training*. [On-Line]. Available:
http://www.dfes.gov.uk/learning&skills/pdf/consultationreport.pdf

DFES (2002b) *Get On with IT - The Post 16 E Learning Strategy Task Force Report*. London: Department for Education and Skills

DfEE (1997) *Connecting the Learning Society*: National Grid for Learning - the Government's consultation paper. London: Department for Education and Employment

Fryer, R.H. (1997). *Learning for the Twenty-First Century. First Report of the National Advisory Group for Continuing Education and LifeLong Learning*. [On-Line]. Available from: http://www.lifelonglearning.co.uk/nagcell/index.htm

Further Education Funding Council (1999a). *Networking LifeLong Learning: Making It Happen*. [On-Line]. Available:
http://www.lscdata.gov.uk/documents/othercouncilpublications/other_pdf/nlllmit.pdf

Further Education Funding Council (1999b). *ILT Implementation Plan*. [On-Line]. Available: http://www.lscdata.gov.uk/documents/circulars/fefc_pubs/9945.pdf

Higginson, G (1996). *The use of technology to support learning in colleges. Report of the Learning and Technology Committee (Higginson Report)*. London: Further Education Funding Council

Kennedy, H. (1997). *Learning Works: Widening Participation in FE*. London: Further Education Funding Council

National Committee of Inquiry into Higher Education (NCIHE) (1997). *Higher Education in the Learning Society*. [On-Line]. Available:
http://www.leeds.ac.uk/educol/ncihe/

Scottish Office (1998). *Opportunity Scotland. A Paper on Life Long Learning*. Stationery Office.

Tomlinson, J. (1996). Inclusive Learning: Principles and Recommendations. London: Further Education Funding Council.

Notes

[1]Europe in the Round Product Information: http://www.gesvt.com/eitrweb/index.htm

[2]Colleges Open Learning Exchange Group: http://www.coleg.org.uk/

[3]Further Education National Consortium: http://www.fenc.org.uk/

[4] Adapt Project web site http://www.adapt.ecotec.co.uk/index02.htm

[5] Ariadne: http://www.ecotec.com/sharedtetriss/projects/files/ariadne.html

[6]Scottish learning Network: http://www.neurvana.com/sln/posfl/index.html

[7] Virtual Learning Centre: http://www.sfeu.ac.uk/learning.asp?pageID=5.1

[8] The Quilt Project: http://www.ccm.ac.uk/ltech/archive/quilt/default.asp

[9] Sheffield College Lettol web site http://www.sheffcol.ac.uk/lettol/

[10] Scottish Further Education Funding Councils ICT Strategy:
http://www.sfefc.ac.uk/about_us/departments/qli/learning_innovation/ictstrategy.html
[11] Becta ICT Survey in Scotland:
http://www.sfefc.ac.uk/about_us/departments/qli/learning_innovation/bectasurvey.html
[12] X4L: http://www.jisc.ac.uk/index.cfm?name=programme_x4l
[13] ICT in Scottish Further Education: http://www.scotfeict.ac.uk/
[14] National Learning Network: http://www.nln.ac.uk
[15] NLN Evaluation:
http://www.ccm.ac.uk/ltech/nlneval/quirinus_show.asp?target=nln_eval&queryid=3&ID=1
[16] Further Education Resources for Learning web site http://www.ferl.org.uk/
[17] Computers for Teachers in FE scheme: http://cfet.ngfl.gov.uk/CFET.pdf
[18] JISCInfonet: http://www.jiscinfonet.ac.uk
[19] CETIS : http://www.cetis.ac.uk
[20] TechDis: http://www.techdis.ac.uk/
[21] Welsh Video Network: http://www.wvn.ja.net/
[22] NIIMLE: http://www.niimle.ac.uk:
[23] Distributed and Electronic Learning Group (DELG) Report:
http://www.lsc.gov.uk/news_docs/DELG_REP0627.doc
[24] SFEFC Summary of ICT Strategic Innovation Project Deliverables and Post Funding
Support: http://www.sfefc.ac.uk/content/library/circs/circs02/fe3702/fe3702annexb.htm#top
[25]Online for a Better Wales: http://www.cymruarlein.wales.gov.uk/skillict/consultation.htm
[26] http://www.scottish.parliament.uk/official_report/cttee/enter-02/elr02-09-01.htm

6

From Pioneers to Partners: The Changing Voices of Staff Developers

Allison Littlejohn and Susi Peacock

In 1997 the National Committee of Inquiry into Higher Education (NCIHE) articulated a vision that foresaw students connecting to powerful academic networks via small portable personal computers. It was envisaged that learning support environments would allow students flexible access to distributed resources, which would promote collaboration and dialogue. Over this period we have seen significant improvements in infrastructure, connectivity and access to computers. The infrastructure is now largely in place but technology in teaching is not ubiquitous and the vision is far from being realised (Beetham, Jones & Gornall, 2001). This is because technological issues have in the main been easier to solve than the more complex social, cultural and organisational issues involved with mainstreaming technology in learning and teaching. As Kenneth Green, Director of the Campus Computing Project in the United States observed:

> The campus community's major technology challenges involve human factors – assisting students and faculty to make effective use of new technologies in ways that support teaching, learning, instruction and scholarship (Green, 1999).

Major government reviews (for example, NCIHE) have seen staff support as a vital tool to overcome the more complex and messy human factors. Staff development in learning technology has been evolving over the last ten years within higher education in the United Kingdom (UK). In this chapter, we explore past successes and consider how we can extend the effectiveness of staff development by:

- Critically reviewing changes in staff development over the last ten years;
- Drawing out the emerging issues and trends;
- Analysing these and their impact on staff development provision.

This will help us to explore future challenges for staff development.

Learning Technology Staff Development: A Historical Perspective

Whilst reflecting on issues in learning technology in higher education over the last ten years, we identified that there have been a series of transformations in the field, occurring at a frequency of approximately two to three years. Therefore, we can view the progression of learning technology staff development in terms of a timeline comprising a series of 'eras', which we have termed pioneering, practice, policy, pedagogy and partnership.

The pioneering era

The first era, set within the formative years of the Association for Learning Technology (ALT) in the early 1990s, can be described as a pioneering stage. During this period learning technology was in its infancy, largely because staff and students had limited access to computers and networks. This era is characterised by early adopters who were skilled, knowledgeable, interested in technology and extremely self-reliant. These pioneers tended to be academics and this led to their use of technology being heavily contextualised within their individual subject disciplines. This corollary is reflected in reviews of Computer Aided Learning (CAL) packages within the early editions of ALT-J (see for example, Soper & MacDonald 1994; Cates, Fontana & White 1993).

The early pioneers believed that this strong focus on subject content enhanced by being presented in a rich multimedia format would engage and motivate learners. Therefore, many of these packages were developed through funding from the UK's Higher Education Funding Councils, through programmes such as the Teaching and Learning Technology Programme (TLTP) [1]. The production of these CAL packages usually required specialist knowledge and IT skills and so they were largely created by project teams comprising subject specialists and multimedia developers. As such, these CAL packages were viewed as "finished products", which were not designed to be repurposed by the end user. In addition, the heavy reliance on large multimedia files led to packages being largely CD-ROM based. Although commercially produced CD-ROM based information resources were being increasingly used within schools and further education (see Chapter 5), there was not widespread uptake by staff within higher education (O'Hagan, 2003). There was, in many ways, a false assumption that exposure to computers and CAL packages was sufficient to drive the development of new forms of teaching with technology. Lack of investment in staff development support in how to use these products resulted in a short lifetime of the courseware.

During the early 90s the norm for staff development in learning and teaching was the one-day or half day workshop, sometimes bringing in outside experts. This was the approach followed for educational technologies like the overhead projector and video and so it was at first for the burgeoning use of the Internet. In 1993 – 94 I remember running events like "A Netsurfing Safari" to help staff learn to use email, Mailbase lists and, of course, how to access early WWW sites. This last was still very new and relatively primitive, and so when it came to authoring learning resources, we urged the use of multimedia software, and even produced a stripped-down authoring shell called TOTAL: Tutor Only Transfer of Authored Learning which was developed by our budding multimedia development unit, in Authorware.

I was still concerned that staff should develop skills in the more conventional pedagogic methods and technologies, considering these a precondition to effective skills in the emerging digital technologies. I provided staff development sessions in student-centred learning, resource-based learning, and open and distance learning.

But I realised that the traditional workshop was inappropriate for in-depth development and embedding. What was needed was staff development itself embedded in real curriculum development projects, particularly where new technology and/or new modes of delivery were concerned. In 1994 the first significant Academic Development Fund was launched in my institution, with £200K over eighteen months. Use of technology was not an obligation, but a climate of expectation had been created and the majority of bids were orientated towards engagement with old and new educational technologies. We could have disbursed three times the sum. The box of pent-up ambitions and ideas was opened, and there was no turning back in terms of such a level of development funding nor in terms of the growing application of technology to learning and teaching.

Figure 1. Chris O'Hagan: A voice from the pioneering era.

The limited number of staff developers in this era were general enthusiasts or IT staff who provided primarily technical support. Staff development workshops generally focused on developing courseware using packages such as Authorware and Macromedia Director. For example, Chris O'Hagan shares his experiences with us as a staff developer in the early 90s (see Figure 1).

In his account, O'Hagan recognises that embedding the use of CAL packages required more than merely demonstrating them in the workshop setting. This approach had been used but had failed to engage academics not already using courseware packages. Furthermore it promoted scepticism about the benefits of computers in learning and teaching.

The practice era

The mid-1990s, saw a dramatic increase in the availability and use of computers. Considerably more staff and students had access to email and the Web. This resulted in greater use and diversity of practice in learning technologies, marking a period which can be described as the Practice era.

In this era, despite an increase in resources, academics were still unlikely to adopt learning technology into teaching practice. To support academics that wished to use technology in teaching, the UK government funded a national programme: The Computers in Technology Initiative (CTI). In order to continue the subject contextualisation, CTI consisted of 24 subject centres distributed across the UK. This significantly increased the profile of technology in teaching by publishing regular newsletters reviewing the latest developments in the use of technology for learning as well as a journal, Active Learning [2]. However, it did little to integrate learning technology with mainstream teaching methodology.

Although the level of support for academics using computers in teaching was growing, it still had limited impact. Further initiatives were instigated by the Funding Councils, including: The Learning Technology Dissemination Initiative (LTDI) [3], TALiSMAN [4] and Netskills [5]. These initiatives provided ideas, support and training, which was integrated into institutions across the UK. Nora Mogey co-ordinator of LTDI (which was then based at Heriot Watt University in Edinburgh) reflects on her experiences (see Figure 2). As Mogey indicates, many of these initiatives were far sighted and staff developers and academics are still using the resources generated today. However, nationally their limit was restricted because they were not linked to policy.

The Learning Technology Dissemination Initiative (LTDI) was established in 1994 with funding from The Scottish Higher Education Funding Council. Its role was to support and encourage staff across higher education in Scotland to make effective use of technology within their teaching.

Those who thought technology was a great idea were generally engaged in projects exploring the potential of learning technology, but most staff were sceptical and a little apprehensive about getting involved.

The LTDI provided staff support through hundreds of workshops for thousands of participants over its five-year life span. Individual support was also taken up by hundreds of academics. LTDI publications were concise but informative, with the hope of making relevant information easy to identify. LTDI's Information Directory was an early attempt to synthesise details and anticipated output from all technology projects. For example, Implementing Learning Technology is not dissimilar in content from books currently on Kogan Page's list and The Evaluation Cookbook[6] (yes it really was inspired by the cookery books on the shelf in my kitchen!) is still evident on many technologists' shelves.

It was certainly too much to hope that LTDI could really spark a fundamental change. I hope those of us involved did help to build the Scottish community of learning technology users that is still evident and a pattern of collaborative and supportive working that many will continue to benefit from over the coming years.

Figure 2. Nora Mogey: A voice from the practice era.

The policy era

1997 saw the instigation of considerable UK government legislation in the educational arena: the policy era. The legislation aimed to extend opportunities for learners through a growth in lifelong learning (Fryer, 1997) forming new partnerships with employers (for example, the University for Industry) [7] and providing a new vision for higher education in the UK (NCIHE, 1997). Three underpinning factors linked these major initiatives. Firstly they all necessitated the extensive and imaginative use of learning technology. Secondly, they required staff to widen their roles, which meant that thirdly, achieving these goals required effective and extensive staff training and support.

NCIHE recognised the need for a professional body that would support staff in planning their professional development. It recommended the establishment of the Institute for Learning and Teaching in Higher Education (ILTHE) and suggested that:

..all new full-time academic staff with teaching responsibilities [were] required to achieve ... membership of ILTHE for the successful completion of probation. (NCIHE, 1997, Recommendation 48).

Membership of ILTHE could be achieved by a portfolio route or by successful completion of a recognised "pathway" (for example, a Postgraduate Certificate in Higher Education). Consequently, a wide range of accredited pathways and postgraduate programmes were quickly established. ILTHE guidelines on embedding learning technology encouraged the development of modules that combined both pedagogy and technology, placing learning technology as an integral method for supporting learning, alongside the more traditional forms of teaching. These programmes were usually offered to staff in addition to the more traditional staff support workshops, broadening the approach to staff development. No longer was the main focus of the staff developer on providing technical support and demonstrating examples of using courseware.

This major shift in staff development was not without significant challenges. Firstly, many learning technology staff developers had limited previous experience of course validation. This was due to the fact that they were from a wide variety of backgrounds (Beetham, Jones & Gornall, 2001). For example, although some had worked as teachers in subject disciplines, many had previously been employed as IT or library support staff. Secondly, most learning technology staff developers had no previous experience in teaching educational theory. In addition, few resources in educational theory were available, so finding suitable course materials proved a major challenge. National initiatives were set up to help alleviate this problem: for example, the Scottish Higher Education Funding Council's (SHEFC) Communication and Information Technology Programme (ScotCIT) [8]. This programme developed resources including case studies and examples of good practice for staff developers as well as ideas and models for supporting the design of online courses encompassing a range of pedagogical methods.

The pedagogy era

From around 1999 there was an increasing emphasis on the design of online courses based on a wide variety of educational models. This included "blended" learning, which integrates online learning with face-to-face interactions. In addition, there was increasingly widespread availability of Virtual Learning Environments (VLE). A VLE can be described as a software environment, which integrates a series of tools. These tools can support a range of teaching and learning activities. A VLE provides opportunities for academic staff to set up online courses without the need for intensive IT training which had previously been required. To support academics in using VLEs, staff developers extended their programmes to include workshops and courses that integrated a range of approaches to course design.

Despite this approach to blended learning, the CTI had remained focused on computer aided learning and did not engage with the non-technological educational developers. Therefore the work of the CTIs was incorporated into a national body which would support a wide variety of learning and teaching approaches: the Learning and Teaching Support Network (LTSN). This led to closer collaboration amongst learning technology staff developers and staff developers who had a non-technological focus.

The importance of connecting different communities of staff is highlighted by Jean Ritchie (see Figure 3), co-ordinator of two major SHEFC initiatives: the Use of Metropolitan Area Networks Initiative (UMI) and then later the SHEFC Communication and Information Technology Programme (ScotCIT). Jean's comments highlight why staff developers gradually found themselves working closely with individuals who were drawn from many different departments across the institution.

The partnership era

One example of a increasing focus on a commitment to partnership was the EFFECTS Project (Beetham, Jones & Gornall, 2001), a national initiative funded by the JISC Committee for Awareness, Liaison and Training Programme (JCALT). This project provided a picture of the relationships amongst support staff and revealed that a wider range of staff than had previously been imagined were involved in the development and support of learning technology. The project listed several categories of staff who were responsible for learning technology staff development, including 'new specialists' (educational developers with a learning technology specialism, educational researchers, technical developers, materials developers, project managers and general learning technologists), 'learning support professionals' (IT support, library staff and C&IT skills professionals) as well as academics and established professionals working at a strategic level. These staff were located within an average of eight different units frequently a library or learning resources unit. This was seen as an obstacle to effective coordination of effort but did not seem to impinge on promoting overall change (Beetham, Jones, & Gornall, 2001).

I worked for SHEFC between 1996 and 2002, first as coordinator of the Use of Metropolitan Area Networks Initiative (UMI) and later as coordinator of the SHEFC Communication and Information Technology Programme (ScotCIT). Both UMI and ScotCIT were notable for funding projects to carry out work in a wide range of topics ranging from infrastructure improvements to development of software tools. ScotCIT in particular had a large element of staff development. In both of the programmes, project leaders and project workers were drawn from a broad mix of backgrounds. My role enabled me to gain an unusual overview of the tensions and misunderstandings that can exist between the different communities:

- communities do not always understand the tasks that other communities carry out;
- communities do not always understand the constraints that other communities face ;
- communities frequently underestimate each others' contributions;
- communities often do not consult each other ;
- communities often do not learn from each other.

One barrier I noted during ScotCIT was that the learning technology and academic communities tended to publish research papers in peer-reviewed journals and relied on these as a means of communication and to share views and innovations. In contrast, while there were journals of relevance to network engineers, many of the innovations underpinning UMI and ScotCIT were not recorded in this way. In addition, conferences that attracted both technical, academic and staff development staff were rare indeed. Yet, there were major overlapping relevances and often discussions based on false assumptions, because of lack of understanding.

How to cure all this is not obvious. All communities need to be represented at the highest level in any communal projects. This may make all concerned feel uncomfortable from time to time; however as representatives come to appreciate each other's contributions, mutual respect should inevitably follow. In Learning Technology today, different communities need to work together to obtain good results. Perhaps a vital role for staff developers is in helping to linking these communities.

Figure 3. Jean Ritchie: A voice from the pedagogy era.

Staff developers adopted different approaches to co-ordinating the effort to "mainstream" learning technology. One keenly explored approach was the American 'Roundtable Methodology'. Susi Peacock and Catriona Kemp from Queen Margaret University College in Edinburgh share their experiences of this approach (see Figure 4).

As we now move into phase, the portal era, technology is increasingly becoming an integral method of linking communities (e.g. RESULTS [10]). Staff developers are now in a very challenging period, which requires them to work as change agents supporting the implementation of key strategies across institutions. Achieving this goal requires close adherence to learning and teaching strategies as well as bridging communities within and between institutions.

The Roundtable Methodology was originally developed by the US-based Teaching, Learning & Technology Group [9] (TLTG) as a means of establishing a representative group drawn from across an institution to progress the large-scale implementation and support of learning technologies. A Roundtable develops recommendations to enhance learning and teaching through uses of technology and particularly aims to improve communication and collaboration amongst its members and across the institution. It does this by bringing together a wide-ranging and relevant group to direct and support learning technology initiatives. This group will usually include:

- Academic staff;
- Technology professionals;
- Administrative staff, including Registry;
- Librarians;
- Staff developers;
- Students.

Roundtables specifically aim to bring together all types of support staff involved with learning and teaching with technology. As the Roundtable, progresses it identifies goals as task areas, establishing working groups consisting both of its own members and asking others to join where appropriate. It combines a strategic approach with grassroots knowledge, concerns and enthusiasm to managing change relating to learning technologies.

Figure 4. Susi Peacock and Catriona Kemp: voices from the partnership era.

Analysis

Our review has revealed a range of issues that have both influenced and driven changes in learning technology staff development within the UK. These include:

- the ubiquity of technology in everyday life;
- changes in attitudes to the use of technology for learning;
- advances in technology;
- the increasing numbers of academic staff experienced in the use of communications technologies;
- the broadening of teaching methodology;
- professionalisation of learning and teaching.

However, the review also raises a number of questions about the effectiveness of staff development which have not yet been fully addressed:

- Have changes in staff development been guided by the needs of teachers and learners?
- Is staff development taking account of changing staff roles?
- Is learning technology embedded within learning and teaching strategies?
- How can we link local and national staff development initiatives?

In this section we will explore these questions by reflecting upon issues and themes in staff development over the past ten years.

Have changes in staff development been guided by the needs of teachers and learners?

An examination of the major themes in the literature produced by the learning technology community over the past ten years reveals a tension between the "pull" of technological advances and the "push" of national initiatives, which focuses on student learning. The "pull" of technological advances was particularly evident in the early eras. At that time, programmes and workshops supported academics in "how to use" the latest technology. However, they did not, to an extent, encourage the questioning of whether or not the use of any particular software tool in itself benefited learning. That is not to say that underpinning educational models were not taken into consideration, indeed there is plenty of evidence to show that, increasingly, they were. But it was also evident that technologies, such as VLEs, had been implemented despite their deficiencies in being able to support a wide variety of educational models. Although most staff development programmes aimed to improve student learning, these have frequently focused too heavily upon outputs, such as course materials and resources (Oliver, 2002). This was particularly evident during the pioneering era when CD-ROMs were prevalent and was then accentuated with the introduction of VLEs (the pedagogy era), which led to a renewed focus on content delivery (see Chapter 2).

National policy started to redress the balance and "pushed" the focus towards the needs of teachers and learners. For example, the setting up of ILTHE, led to the development of courses, which attempted to satisfy the needs of academics. This had the additional benefit of integrating learning technology within mainstream learning and teaching staff development. While the rewards for innovative teaching provided through professional accreditation are a welcome change, they are somewhat limited in their impact. Staff developers have been struggling to provide on-going incentives for academics to adopt learning technology into their mainstream teaching, for example, increased productivity in terms of content resource production. The result has been a compromise between an ideal implementation of learning technology, which focuses on the benefits for learners, and the realities of working within real organisations with resource and social constraints.

Is staff development taking account of changing staff roles?

> The growing emphasis on learning rather than teaching in higher education means that students can be expected to place increasing demands upon support staff to provide them with advice and guidance…. The task of 'teaching students how to learn' was one they had previously seen as being the responsibility of academics. (NCIHE, 1997 14.10)

Current staff development programmes largely concentrate on academic staff. However, support roles are also changing and staff in these areas require re-skilling to undertake new tasks. For academics, increased use of learning technology means that some staff will gain new responsibilities (such as moderating online discussions). At the same time, they will lose former tasks (such as assuming full control of the learning environment). Letting go of familiar, comfortable practice and adopting new ways of doing things is a painful process, which all too often is ignored (Marris, 1986; Noble, 1998; Taylor, 1999). Many staff development programmes provide support within the cognitive and psychomotor domains (Bloom, 1956). However, programmes also need to reflect upon and take account of academics' anxieties to enable development within the affective domain.

Finally, different staff groups have disparate views of learning technology, as evidenced in Jean Ritchie's case study. We have to find some way of working towards a common vision whilst acknowledging the different staff perspectives. The 'Roundtable' approach has encouraged this kind of holistic approach in which groups of staff interact and debate their individual understandings of learning technology. This team approach ensures that individual groups of staff have a clear understanding of each other's views and values.

Are we adopting coherent approaches to embedding learning technology?

Many institutions view e-learning strategies as being distinct from learning and teaching strategies. This fragmented approach can lead to confusion as to the role of technology in the support of learning. Related to this, a large number of institutions have restructured central support units, resulting in a bewildering mixture of approaches: from centralised 'learning service', 'teaching and learning' and 'academic practice' units to decentralised support within faculties (Alexander, 1999; Beetham, Jones & Gornall, 2001). A further complication is that some units are academic, while others are purely service departments (audio visual, IT and library services). Furthermore, a wide range of staff are working in learning technology and the term learning technologist is very broad. This has caused some difficulty for staff developers who are trying to find a focus for their efforts when catering for someone teaching I.T. skills in how to use a VLE or an educational researcher. This can lead to further uncertainty about the focus of staff development. According to Ehrmann (1998) all education is local and we need to recognise that every institution has individual requirements.

By focusing upon commonalities rather than differences we can explore a common vision for learning technology across communities, rather than trying to force change upon the communities themselves. This common vision may be enhanced through linking national and local agendas.

How can we link local and national agendas?

National staff development initiatives have had limited impact at an institutional level. This is especially the case with short-term projects after funding has ceased (see Chapter 4). Funding bodies expect that project outcomes will be easily embedded within institutions. However, there has been limited institutional support for staff developers in achieving this goal. This could reflect different approaches to collaboration: externally funded projects are frequently based upon the premise that collaboration amongst institutions is mutually beneficial. In comparison, institutions in general have a more introspective and competitive culture, which does not always facilitate collaboration.

Linking national and local initiatives could be achieved by staff developers being actively involved in the planning of learning and teaching strategies. This will ensure that the good practice arising from the outcomes of national projects is linked to local initiatives (McCarten & Hare, 1996). The newly proposed Academy for the Advancement of Learning and Teaching in Higher Education could help strengthen these links and promote collaboration between learning technology staff developers and their non-technological counterparts. This would help bring about a more coherent, integrated and effective approach to staff development.

The Next Ten Years?

By reflecting over the last ten years, we can consider how future staff development can be improved. There is plenty of evidence in the literature that effective staff development must be positioned at multiple levels within an institution. For example, Carmel McNaught (2002) states that:

> Staff development can no longer be a pleasant cottage industry on the fringes of academe...Effective staff development is positioned at the centre of the university function and yet needs to retain connections with the needs and perceptions of teaching staff. This is a demanding challenge.

Both McNnaught et al (1999) and Roberts et al (2002) advocate an approach that is top-down (guided by policy), bottom-up (fuelled by the demands of staff and students) and middle-out (connecting staff at varying levels). These three provide a powerful framework for supporting and embedding learning technology. However, all three are seldom simultaneously implemented within an institution.

In the early pioneering days, enthusiasts provided the bottom-up approach. In the policy era, national and local policies were devised as a top-down driver. It is only since the recent partnership era that middle-out approaches have been implemented. A current challenge for staff developers lies in developing this middle-out approach by connecting communities (for example, academics and other support staff) who are working towards strategic goals. Future challenges lie in combining all three approaches.

The software and hardware we use today will be unrecognisable in ten years with an increase in wireless, portable and even wearable technologies (Van Dam 2002). It is, however, the underpinning processes of learning and teaching that are more likely to endure particularly those involved in connecting communities.

References

Alexander, W. (1999). *Talisman review of staff development courses and materials for ICT in teaching, learning and assessment.* [On-Line]. Available: http://www.talisman.hw.ac.uk/CITreview/cit_index.html

Beetham, H., Jones, S., & Gornall, L. (2001). *Career Development of Learning Technology Staff: Scoping Study Final Report.* [On-line]. Available: http://sh.plym.ac.uk/eds/effects/jcalt-project/final_report_v8.doc

Bloom, B.S. (1956). *Taxonomy of educational objectives: The classification of educational goals: Handbook I, cognitive domain.* New York: Longmans.

Cates, W., Fontana, L., & White, C. (1993). Designing an Interactive Multimedia Instructional Environment: The Civil War Interactive. *ALT-J*, 1, 2, 5 - 16

Ehrmann, S. (1998). *Studying, Teaching, Learning and Technology: a Toolkit from the Flashlight Programme*. [On-Line]. Available: http://www.ilt.ac.uk/public/cti/ActiveLearning/al9pdf/ehrmann.pdf

Fryer, R.H. (1997). *Learning for the 21st Century*. National Advisory Group for Continuing Education and Lifelong Learning. [On-Line]. Available: http://www.lifelonglearning.co.uk/nagcell/

Green, K.C. (1999). *The Continuing Challenge of Instructional Integration and User Support. The Campus Computing Project*. [On-Line]. Available: http://www.campuscomputing.net/summaries/1999/

McCartan, A., & Hare, C. (1996). Effecting Institutional Change: the Impact of Strategic Issues on the Use of IT. *ALT-J*, 4, 3, 21-8.

McNaught, C., Kenny, J., Kennedy, P., & Lord, R. (1999). Developing and Evaluating a University-wide Online Distributed Learning System: The Experience at RMIT University. *Educational Technology & Society* 2, 4. [On-Line]. Available: *http://ifets.ieee.org/periodical/vol_4_99/mcnaught.html*

McNaught, C. (2002). Views on staff development for networked learning, In C. Steeples & C. Jones, (Eds.) *Networked learning: perspectives and issues* (pp111-124). London: Springer.

Marris, P. (1986). *Loss and Change*. London: Routledge

National Committee of Inquiry into Higher Education (NCIHE) (1997) *Higher Education in the Learning Society*. [On-Line]. Available: http://www.leeds.ac.uk/educol/ncihe/

Noble, D. (1998). Digital Diploma Mills: The Automation of Higher Education. *First Monday*, 3.1. [On-Line]. Available: http://www.firstmonday.dk/issues/issue3_1/noble/

O'Hagan, C. (2003). Implementing Learning Technologies at the University of Derby, 1989-2003: A Case Study. *The Observatory on Borderless Higher Education*, 13.

Oliver, M. (2002). *Metadata vs. educational culture: roles, power and standardisation*. Paper presented at Ideas in Cyberspace Conference, Higham Hall, England.

Ritchie, J. (2002). *ScotCIT final report: Achievements, Outcomes, Recommendations*. [On-Line]. Available: http://www.shefc.ac.uk/about_us/departments/qli/scotcit.pdf.

Roberts, J., Brindely, J., Mugridge, I., & Howard, J. (2002). Faculty and staff development in higher education: the key to using ICT appropriately? *The Observatory on Borderless Higher Education*, 12

Soper, J., & MacDonald, A. (1993). An Interactive Approach to Learning Economics: The WinEcon Package *ALT-J*, 2, 1, 14-29

Taylor, P. (1999). *Making sense of academic life: Academics, Universities and Change*. Buckingham: Open University/SRHE press.

Van Dam, A (2002) Next Generation Educational Software. In *Proceedings of the World Conference on Educational Multimedia, Hypermedia and Telecommunications*, Vol. 2002, Issue. 1.

Notes

[1] Projects within the Teaching and Learning Technology Programme included: Mathwise, http://www.bham.ac.uk/mathwise/, STEPS http://www.stats.gla.ac.uk/steps/, StoMP, http://www.ph.surrey.ac.uk/stomp/, and Computer Aided Learning for Pharmaceutical and Life Sciences, http://www.coacs.com/PCCAL/ and Geography, http://www.geog.le.ac.uk/cti/tltp/.

[2] An archive of Active Learning is available at:
http://www.ilt.ac.uk/public/cti/ActiveLearning/index.html

[3] LTDI: http://www.icbl.hw.ac.uk/ltdi/

[4] Teaching and Learning in Scotland over the Metropolitan Area Networks (TALiSMAN):
http://www.talisman.hw.ac.uk/

[5] Netskills: http://www.netskills.ac.uk

[6] The LTDI Cookbook is available at: http://www.icbl.hw.ac.uk/ltdi/cookbook/contents.html

[7] University for Industry: http://www.dfee.gov.uk/ufi/

[8] ScotCIT: http://www.scotcit.ac.uk

[9] The TLTG: http://www.tltgroup.org

[10] RESULTs: http://www.results.ac.uk

7

Institutional Implementation of ICT in Higher Education: A Dutch Perspective

Wim de Boer, Petra Boezerooy and Petra Fisser

The environment in which higher education institutions have to operate has changed significantly in the last decade and is still changing. New student populations, governmental policy, new conceptions of learning and many other changes related to technological developments are causing higher education institutions to rethink their way of teaching and doing research. In the Netherlands these changes can also be seen. This chapter will discuss the changing context in Dutch higher education and describe the current experiences of teaching and learning with Information and Communications Technologies (ICT). Learning models and how these relate to the use of ICT in higher education will be discussed and conclusions about the current and future use of ICT in Dutch higher education institutions will be drawn.

Factors That Influence Changes in Higher Education

In research about changes in higher education it can be seen that the changes most important to higher education institutions are initiated externally (e.g. Levine, 2000; Bates, 2001; Fisser, 2001; Middlehurst, 2003). Examples of these external influences are governmental and policy developments, demographic changes, market forces, the knowledge economy, internationalisation of higher education and lifelong learning. Furthermore, it is well known that higher education institutions have to react to these external changes in order to survive. The use of ICT is seen as one

of the responses to these external changes. ICT has enlarged the opportunities available for higher education organisations and is having an effect on the traditional
modes of production, ways of communicating with students, organizational structures, budgeting and accountability mechanisms and quality assurance procedures.
However, as Fisser (2001) shows in her overview framework of factors that could
have an impact on the use of ICT in education, alongside the external influences,
institutional conditions could have an influence on the use of ICT. Fisser (2001)
categorised factors that effect using new forms of ICT in education into six groups,
listed in Table 1.

An important issue to consider is how individual Dutch higher education institutions adapt to the external and internal developments. Do they use ICT for reducing
geographical distances in order to make these distances less of a barrier for both
students and higher education institutions? Or do they change the roles of instructors
or make use of more flexible curricula and new teaching and learning paradigms? In
relation to this, one can wonder whether Dutch higher education institutions are in
the process of providing quality education for rapidly diversifying student cohorts
(Middlehurst, 2003), for which the ideal mode of delivery is a mix of on-campus and
flexible learning (Bates, 2001). These are some of the questions we will address in
this chapter. We will not only look into the current situation of the use of ICT in
Dutch higher education, but will also focus on some of the future perspectives.

Table 1. Summary of factors that effect using new forms of ICT in education.

Category	Factors	
Environmental pressures	New market	Competition
	Education as business	Response to threats and
	Part-time students	opportunities
	Lifelong learning	Flexibility
	On-demand training	Knowledge management
	Funding	Changing student demo-
	Partnerships	graphics
	Tailor-made products	Demands from employers
	Dynamic environment	Demands from learners
Technology developments	Emerging technology	New technology (push,
	Dependence on IT	hype)
Institutional conditions	New organisational struc-	Concrete plans
	ture	Improved access to educa-
	Broad participation	tion Leadership
	Shared vision	
Educational develop-	New conceptions of learn-	Individual differences
ments	ing	Active learning
	New teaching models	Focus on learner/learning
Cost reduction / Cost-	Reducing costs	Benefits
effectiveness	Cost-effectiveness	
Support facilities	Administrative support	Availability of technology
	Educational and technical	Availability of facilities
	support (including staff	
	development)	

Drivers for Change for Dutch Higher Education Institutions

The Dutch higher education system is a binary or dual system, consisting of both universities and universities of professional education. Alongside these two major sectors, higher education is also provided through the Dutch Open University. The Dutch higher education system typically combines a centralised governmental policy with a decentralised policy (institutional autonomy) for the administration and management of the higher education institutions. With respect to the use of ICT this means that, contrary to the existence of some central governmental ICT policy for the primary and secondary education sector, governmental ICT policy for the higher education sector hardly exists. However, there are some semi-governmental funded initiatives that can be seen as the main drivers for change with respect to the use of ICT within Dutch higher education. These two initiatives are the SURF Foundation and the Dutch Digital University, two organisations that promote the use of ICT in higher education.

The SURF Foundation

The mission of the SURF Foundation is to exploit and improve a common advanced ICT infrastructure that will enable higher education institutes to better realise their own ambitions and improve the quality of learning, teaching and research (Stichting SURF, 2003). The SURF activities are funded by the higher education institutes in the Netherlands as well as by the Dutch government. One of the programmes within the SURF Foundation is the SURF ICT and Education Platform. This platform has generated a wealth of experience and products and paved the way for further cultural changes. Building on four years of successful tenders for innovation projects, the SURF ICT and Education Platform will support innovative projects during the next few years. Each year one or more themes will be adopted, in which modernisation projects that complement and strengthen each other will be subsidised and implemented via a tender. Institutions in collaboration with each other write the project plans within this tender. The platform itself initiates various projects, which promote, for example, systems integration or putting in place the use of standards.

The Digital University

The Dutch Digital University is a consortium of ten higher education institutions in the Netherlands. It focuses on the development and application of digital educational products and knowledge for higher education (Digitale Universiteit, 2002). Important issues for the Dutch Digital University are a changing demand for education, combining working and learning, permanent education, the role of e-learning and the need for cooperation. The Dutch Digital University aims to set up a relevant knowledge network, share expertise and, last but not least, share the financial burden of innovation. The projects of the Dutch Digital University can be divided into five programs:
- Digital testing, assessments and digital portfolio;
- Digital educational tools: tasks and resources;
- Learning and coaching from a distance: dual, virtual and international;
- Building up and disseminating expertise;

- Electronic Learning environments (standardization and interoperability).

In addition to the activities of the SURF Foundation and the Dutch Digital University, many projects have been set up within the individual higher education institutions. How these and other initiatives have influenced the use of ICT in higher education will be discussed in the next section.

Current Experiences of Teaching and Learning With ICT

In describing the Dutch situation, with respect to the use of ICT, we will rely heavily on the work of Collis and Van der Wende (2002) who conducted an international comparative study of 'Models of Technology and Change in Higher Education' in the Netherlands, United Kingdom (UK) and Australia. An overview will be given of the results for Dutch higher education (more detailed information can be found in De Boer and Boezerooy 2003).

In the last decade, at all levels in Dutch higher education institutions, innovative ICT experiments have been conducted. Many of these institutions have expanded the pioneering stage through the so-called '1,000 flowers blooming' phase, to faculty and even institution wide managed change or the so-called 'bottom-up to top-down approach' (Fisser, 2001). An important condition for innovative use is a high level of technical infrastructure. The technical infrastructure of the Dutch higher education institutions (supplied via SURFnet, part of the SURF Foundation) is one of the world's fastest and most advanced networks. Speed, reliability and security of the network are the key issues. De Boer and Boezerooy (2003) note that an estimated 400.000 staff and students of over 200 organisations (including the Dutch universities, universities of professional education, academic hospitals, research centres and (scientific) libraries) are connected to SURFnet. Students and staff of higher education institutions can have access to SURFnet from both the office and at home.

Another important factor for fostering innovation is institutional policy. There are many external or environmental pressures and institutional conditions that can have an effect on the institutional policy with respect to the use of ICT. These pressures and conditions are highlighted in the definitions and descriptions given by Collis and Gommer (2000) and Collis and Moonen (2001) of four main scenarios for educational delivery that can have an influence on the institutional policy (see Figure 1). The scenarios are situated within two dimensions (or lines of change). The first dimension relates to the extent to which institutions focus on local or global issues. The second dimension relates to the extent to which the institution or the lecturer controls the quality of the program.

In Scenario A (Back to Basics), the higher education institution offers on-campus activities, in which the institution controls the program. Furthermore, there is hardly any flexibility for the 18-24 years old on-campus students. However, in this scenario it is also the case that many higher education institutions are starting to experiment with distance participation in their established programs. This can lead to Scenario B (The Global Campus). In this scenario the institution also controls the program, but there is a focus on more diversified off-campus student groups. Scenario C (Stretching the Mould) relates to increased flexibility with or without

Scenarios	*Where local and face-to-face transactions are highly valued*	*Where local and face-to-face transactions are highly valued*
In which the institution offers a program and ensures its quality	**SCENARIO A** Quality Control of a cohesive curriculum, experienced in the local setting. **Back to Basics**	**Scenario B** Quality Control of a cohesive local curriculum, available globally. **The Global Campus**
In which the institution offers a program and ensures its quality	Scenario C Individualisation in the local institution. **Stretching the Mould**	Scenario D Individualisation and globalisation. **The New Economy**

Figure 1. Four scenarios for educational delivery (Collis & Gommer: 2000: 32).

changing the underlying pedagogical model within the institution. Furthermore, institutions offer more flexibility for participation within their pre-set on-campus programs. Scenario D, The New Economy, is the most radical; a student can make his own decisions about what, when, how, where, and with whom he learns.

De Boer and Boezerooy (2003) used these four scenarios as a framework for describing the current use of ICT in Dutch higher education institutions. They found that Dutch higher education institutions mainly focus on the traditional on-campus activities, in which face-to-face contact and contact between the instructor and students are two of the most important aspects. At the same time there is less emphasis on offering time and place independent learning for a diversified target group. In Dutch higher education, teaching and learning with ICT is mainly aimed at the 18-24 years old students and there is far less emphasis on international students and lifelong learners. In addition to the emphasis on offering time and place independent learning, Dutch higher education institutions only offer a moderate choice in the programs they offer. The institutions decide upon the programs they offer and these programs are in principle fully planned, with some individual choices for students. Institutions offering highly flexible programs in which students can choose more or less their own combinations are rare inside the Netherlands.

With respect to the use of ICT in teaching and learning De Boer and Boezerooy (2003) found that technology, in terms of e-mail, word processing, PowerPoint and Internet has become standard as part of the teaching and learning process. This trend can also be seen for the implementation and use of the electronic or virtual learning environments (VLE) in Dutch higher education. Almost all of the Dutch higher education institutions have implemented or are implementing an electronic learning environment. The most popular systems are Blackboard, WebCT and Lotus Learning Space, as well as home made systems such as TeleTOP, Polaris and N@Tschool.

One of these electronic learning environments, TeleTOP, has been developed at the University of Twente. An analysis of the use of TeleTOP was undertaken by De Boer (2003) and from this analysis it appears that most faculties at the university have implemented TeleTOP. The implementation of course environments in Tele-TOP started most commonly with first year courses, followed in the next year by second year courses and so on. In the academic years 2000/2001 – 2002/2003, 2766 TeleTOP course environments were set-up at the university of Twente. Of these TeleTOP environments, 83% (2268) were produced for courses, the other environments were used as project environments. The average number of TeleTOP course environments that are being produced has been about a 1000 per year. Furthermore, the analysis showed that 73% of the 2268 course environments were actually used for course support, in which the instructor was responsible for the content (teaching and learning materials) and the students had access to the environments. A minimal requirement for the use of TeleTOP is that for each course environment at least five documents had to be placed in the environment. The average number of the Tele-TOP documents that an instructor placed to the environment is around 105 documents. It is interesting to note that the larger the student groups, the more documents an instructor placed in TeleTOP.

Although standard applications and electronic learning environments such as TeleTOP have become a more common phenomenon in the teaching and learning process they have not radically affected the nature of this process. The instructor is and will remain the "core medium". This indicates that the classroom orientation model is the most common model used within Dutch higher education; a model in which instructors and other actors highly value the face-to-face interaction and direct communication between instructors and students and among students (see Table 2 for results of a survey of decision-makers, support staff and lecturers). However,

Table 2. Use of technology in Dutch higher education: Part of a blend.

Features	Scale (1-5)	Mean (N=57)	SD
How much interaction with the instructor occurs in the course?	Very low amount-Very high amount	3.30	.71
How much interaction among the students occurs in the course?	Very low amount- Very high amount	3.32	.66
How are the learning materials used in the course acquired??	All predefined/ acquired by the instructor- All found or created by the students	2.91	.58
How does the student participate in the course?	Individually- As part of a group	2.91	.61
How much of the course is web-based?	None- Entire course is web-based	2.89	.72
How does the student communicate within the course?	Face to face- Only via computer	2.88	.38

the more flexible ways of learning, such as communication via the Web or more flexibility for students in choosing their teaching and learning materials, are gaining interest in Dutch higher education. They do not replace the traditional on-campus settings, but complement them and become part of the blend of on-campus delivery.

Future Expectations for Learning Models and Related ICT: From Back-to-the Basics to Stretching The Mould?

In the precious section, four different scenarios for educational delivery were out-lined. De Boer and Boezerooy (2003) report on results of a survey where decision makers, support staff and lecturers in Dutch higher education institutions were asked to indicate the extent to which these four scenarios exist now and the extent to which they would exist in the future (2005). The results, as outlined in Table 3, reveal that on-campus activities for the 18-24 years old students dominate both current and fu-ture descriptions of practice.

Whilst face-to-face contact with the traditional (18-24 year old) student groups will remain important in the future, ICT will become increasingly part of the blend of technology and traditional ways of teaching and learning. Furthermore, no real dramatic changes in mission, profile or market position are expected, especially not with respect to new target groups like international students and lifelong learners. Nevertheless, Dutch institutions are gradually "stretching the mould"; offering more flexibility in changing their procedures, models and programs as a process of change from within. It seems that within courses more flexibility is going to be offered. These changes, however, are gradual and usually slow and may comply with the slight changes in needs and demands as perceived by the institutions.

Table 3. Extent to which typical learning settings occur now and in the future (DeBoer and Boezerooy, 2003).

Scenario	Typical learning set-ting (N=57)	Now	Future
		Mean (SD)	Mean (SD)
A	Back to the basics	4.55 (0.75)	4.23 (0.82)
B	The global campus	1.70 (0.78)	2.76 (1.03)
C	Stretching the mould	3.50 (1.07)	4.14 (0.79)
D	The new economy	1.52 (0.78)	2.70 (1.12)

1=little or none, 3=some, 5=very much the case

It is interesting to compare these results with those of some of the other countries involved in the international survey (Collis & Van de Wende, 2002). Results from this comparison indicate that for UK and Australian higher education institutions the most common scenario for describing their current situation is the "Back to Basics" scenario. Predictions for the future however, reveal a move towards widening the opportunities for distance learning. This opening and stretching of the traditional course model (De Boer & Collis, 2003) seems a way for actors within the higher education to meet demands such as lifelong learning and providing programs for international students. However, Dutch higher education institutions are far less concerned with meeting the demands of international students than their UK and Australian counterparts (See Chapters 8 & 9).

Conclusion

As in many other countries, the impact of ICT in Dutch higher education has been considerable, but also very diverse. In the 80s and the 90s many experiments were started within the individual higher education institutions and many of these experiments have become institutionalised. However, this does not (yet) mean that the introduction and development of ICT has had the wide-ranging effect on the teaching and learning processes in Dutch higher education institutions that was expected or predicted by many people.

With respect to the near future, De Boer & Boezerooy (2003) report that Dutch higher education institutions do not expect any revolutionary change as a result from or related to the use of ICT. There is not really a concern about being forced to change by either external forces or institutional developments. But these factors are having an influence. With a strong institutional policy and important key actors that promote the use of ICT in education, more ICT initiatives are becoming institutionalised. Even though ICT in education has promising possibilities in relation to distance learning and online learning, campus-based variations, which offer more flexibility, will be a primary focus both now and in the future. Nevertheless, modest changes will occur in relation to distance learning, but only parallel to the on-campus mode, not replacing it.

References

Bates, T. (2001) *National strategies for e-learning in post-secondary education and training*. Paris: UNESCO/IIEP.

Boer, W. F. de, & Collis, B. (2003). Flexibility beyond time and place: Stretching and opening the course. In A. Szucs & E. Wagner (eds.). Conference Proceedings of EDEN 2002: The Quality Dialogue (pp. 95-101). Bulgaria: EDEN.

Boer W.F de (2003). *TeleTOP evaluation*. Internal report. Enschede: University of Twente.

Boer, W.F. de, & P. Boezerooy (2003). ICT in the Netherlands: Current experiences with ICT in higher education, in: M. van der Wende & M. van der Ven (Eds.) *The Use of ICT in Higher Education: A Mirror of Europe*. Utrecht: Lemma.

Collis, B., & E. M. Gommer (2000). Stretching the Mold or a New Economy? Scenarios for the university in 2005. *Educational Technology*, XLI (3), pp. 5-18.

Collis, B., & J. Moonen (2001). *Flexible learning in a digital world: Experiences and expectations*. London: Kogan Page.

Collis, B., & M. van der Wende (2002). *Models of technology and change in higher education: An international comparative survey on the current and future use of ICT in higher education*. Enschede: CHEPS.

Digitale Universiteit (2002). *Strategisch samenwerken aan vernieuwing* (Strategic cooperation for educational change). Utrecht: Stichting Digitale Universiteit.

Fisser, P. (2001). *Using Information and Communication Technology, a process of change in higher education*. Enschede: Twente University Press.

Levine, A.E. (2000). *The future of colleges: 9 inevitable changes. The Chronicle Review, Chronicle of Higher Education*, 47: B10-B11.

Middlehurst R (2003). Competition, Collaboration and ICT: Challenges and Choices for Higher Education Institutions. In: M. van der Wende & M. van der Ven (Eds.) *The Use of ICT in Higher Education: A Mirror of Europe*. Utrecht: Lemma, 2003.

Stichting SURF (2003). *SURF Jaarplan 2003, ICT en Onderwijs*, Utrecht: Stichting SURF.

Wende, M., van der & M. van der Ven (2003). *The use of ICT in higher education; A Mirror of Europe*. Utrecht: Lemma.

8

Institutional Implementation of ICT in Higher Education: an Australian Perspective

Ron Oliver, John O'Donoghue and Barry Harper

Australia is a country that has a long history of providing flexible learning opportunities for students in higher education. It has a distance education record that goes back further than most other countries and is one of the countries where quality in teaching and learning has been an item on the Governmental agenda for some time. Currently Australia is a country with a high degree of expertise and capability among its university sector in the use of Information Communication Technologies (ICT). Tracing the history of the uptake of ICT in higher education provides some insights that reveal the various influences that have led the sector to its current position. At the same time, a study of the history suggests a number of issues that could be addressed to advance these activities even further.

We have seen considerable change in the way ICT has been used in higher education with time (Singh, O'Donoghue & Betts, 2002). If we exclude the use of ICT as an administrative tool for managing and organising education, there have been a number of stages that appear to describe the use of ICT. Prior to 1990, the use of ICT as an instructional tool in any Australian university was likely to be as a consequence of the dedicated and often individual efforts of early technology adaptors seeking learning gains within their own teaching. More recently ICT has become an agent and transformer supporting moves to flexible delivery. The purpose of this chapter is to trace the development of ICT in higher education in the Australian context and to map some of the significant events along the way.

Technology and Distance Education

The use of technology in higher education as an instructional tool tends to have been rooted in activities associated with the provision of distance education. In most countries these activities have been well documented and historical accounts trace the use of technology through discrete stages. Garrison (1985) describes these three generations in terms of the technological innovations underpinning each: correspondence, telecommunications and computers. Nipper (1989) describes three similar generations as, correspondence, multi-media and telelearning and in this description separates the media from the instructional methodologies employed. Each of the generations has resulted in a paradigm shift in the delivery mode.

In descriptions of technology use in distance education, early discussions centred on such important concepts as learner independence (e.g. Taylor, 1992 & Catchpole, 1993) and interactivity (e.g. Moore, 1989; Catchpole, 1992). The emerging information technologies in the 1980s provided considerable scope for innovative designers to provide a means for supporting distance learners through enhanced forms of interaction in the materials. The applications of technology to distance education have been distinctly influenced by the need and wish to improve the effectiveness of learning by increasing interactivity. High levels of interaction were often achieved through computer-assisted instruction (CAI). At the time many teachers questioned the degree of interactivity provided by the various types of CAI. While many saw interactions with a computer as being of far less value than interactions with a teacher, Garrison (1985) argued that one should consider the interactions with CAI as interactions with a teacher through the computer. He argued that it was not the hardware and technology providing the responses in interactions but rather the learning framework created by the instructional designer, who in this case was the teacher.

The concept of interactivity in distance learning gained prominence in the late 1980s as information and communication technologies started to emerge. Perhaps the single most important technology providing the means to improve the quality of distance and open education has been telecommunications technology. Telecommunications is used to describe the use of electromagnetic channels to transfer and receive information and this generation of distance education delivery involved the use of such technologies as telephone, television, facsimile, audio conferencing and teleconferencing. The essence of the telecommunications applications was their facility to increase interactivity between teacher and student. There were many instances of telecommunications being used in distance education teaching and not all of these were characterised by high levels of interactivity. For example, television and radio broadcasts typically delivered instruction with no prospect of learner interaction, although there were many instances of these applications being trialled with interactive elements. (e.g. Oliver & McLoughlin, 1997).

Computer-mediated communications quickly followed and were employed in distance education to provide an interactive element to the teaching and learning process. In computer-mediated communications, the teacher and learner are able to interact through computer applications such as electronic mail (email) and telematics. These interactions could be instantaneous or delayed and an array of synchronous or synchronous applications emerged to support learning programs. At that time, many experimental programs were being developed. Hedberg & Harper (1996) reported on a series of case studies and projects, which illustrated the beginning of institutions starting to expand their approaches to open learning through communications, particularly in teacher education and professional development. The experiments described ranged from simple, cheap solutions to quite novel and expensive infrastructures.

Very quickly the affordances of technology led to explorations of flexibility in delivery formats, flexibility being an inherent characteristic in distance education. In the early 90s there was considerable hype in Australia surrounding the notion of open learning. This was driven to a large extent by a social political agenda, with particular reference to disadvantaged groups (Dekkers & Andrews, 2000). A number of initiatives in Australia in this period saw technology of various forms being implemented not only to foster enhanced learning but also to create greater accessibility to courses and programs. The Open Learning Agency is a good example of this [1].

However, what is possible is not always necessarily desirable. Hutchison (1998) warns against getting too tied up in the technology. This prompts a consideration of whether online education or the virtual university merely offers instruction and not a real education that a traditional university can offer where students hone their communication skills, mix with people from diverse races and cultures and increase their understanding and empathy of those different races. Baumbach and Bird (1998) take the practice and philosophy of a "good educator" and consider "technologising" it. This raises significant issues about whether we want to simply interweave existing practice with technology or to develop a new modus operandi utilising technology (as suggested by the University College of the Cariboo's IT plan, developed in 1998).

Convergence

In the early 1990s the developments driving technology applications in distance learning settings started to merge with developments driving technology usage in classroom and face-to-face teaching. Evans and Nation (1993) provide a description of the forces that were driving this convergence between distance-education and on-campus teaching and learning methods. When one examines the weaknesses of on-campus teaching and learning, there were then, and still are today, many sound reasons for use of the new technologies, for example, empowering learners and lessening their reliance on teachers and instructors.

Several studies at the time when resource and technology-based educational programmes were being implemented revealed poor acceptance on the part of many of the stakeholders. Kelly, for example, (1987) describes a project where an on-campus programme was modified from the conventional mode to a mode that provided students with less lecturing and guidance and placed resources and materials at the disposal of students to enable a student-centred and self-paced approach to learning.

Responses from the students to this approach were less than favourable and despite repeated modifications and alterations, full acceptance of the approach and the ideology behind it was never gained. A research base that might have guided the teachers and trainers in this area never really emerged (see Chapter 10 for a fuller discussion of the development and rigour of learning technology research). Where research was being conducted into the effectiveness of the technology, many of the methods and activities were less than rigorous and not able to provide the answers that were required (Reeves, 1993).

In descriptions of teaching and learning activities at Griffith University, where instructional methodologies from distance education programmes were employed in on-campus teaching, Ross (1993) describes difficulties when not all parties perceived a problem to exist. The project involved the use of multimedia products and distance education packages to provide student-centred learning and the replacement of some forms of traditional teaching with technology based alternatives. Some of the difficulties experienced in this setting included poor student response, high costs and a perceived erosion of conditions for both staff and students.

Dunnett (1990), in describing the use of technology in education, compared and contrasted applications where the media was supportive of the instruction and instances where the media was essential to the instruction. He saw a significant distinction when the same media were applied across distance and on-campus education because the media necessarily served disparate purposes in each. Moves at that time in the on-campus circles of higher education within Australia to adopt new technologies to supplement on-campus teaching showed symptoms of what is called "technology-led curriculum development". This model of adoption of new technologies into education is still practised in systems worldwide (Reeves, 1992) and many questions remain in terms of its effectiveness and utility.

The whole question of the use of media in the learning process was the subject of considerable debate in academic circles at the time. While many educators were rushing to embrace applications of the new technologies as multimedia and computer managed learning, they were counselled to be cautious in their expectations and anticipations (e.g. Schramm, 1977; Clark, 1983; Clark & Salomon, 1986). Research appeared to indicate that media themselves did not influence learning, but rather it was the instructional design accompanying the media that influences the quality and quantity of learning. Clark argued this with the observation that the media (delivery technologies) are "mere vehicles that deliver instruction but do not influence student achievement any more than the truck that delivers our groceries causes changes in our nutrition" (Clark ,1983:455). However there were strong opposing views (e.g. Cuban, 1986, Kozma, 1991).

Flexible Delivery

Emerging from the growth of ICT in distance education was a concerted drive to adopt these technologies to support enhanced forms of learning and modes of flexible delivery. Nunan (1996) describes flexible delivery as " the process of increasing flexibility in learning" and includes such elements as time, place and mode among the flexibilities to be achieved. Quality online learning is not simply a question of putting existing lecture notes onto the Internet. It requires special skills to identify what multimedia method would be the best for delivery of an online course. Should the teacher use a group discussion via videoconferencing or simply accept basic email messages from students and be swamped by emails? These decisions may only be made with experience and after extensive investment by institutions in staff training.

The adoption of technology as an aid to teaching and learning in education has been a long process that still seems to be a considerable way from its potential end-point. Moves to flexible delivery strategies incorporating new technologies have been based on a number of lures and expectations. For example:

- The capacity to build and support virtual learning communities (e.g. Flexible Learning and Higher Education Resources [2]);
- The development of new ICT and capabilities;
- The development of lifelong learning and the notion of a learning individual as important outcomes from university education (e.g. Trevitt, 2000);
- The development of a learning profession and a learning discipline as a concept intermediate between the notion of a learning individual and a learning organi-sation. (e.g. Trevitt, 2000);
- Funding reductions and the transition to a commercialised globalised university ethos (e.g. Coaldrake & Steadman, 1998);
- The potential for greater cost efficiency in the delivery of courses and programs (e.g. Inglis, Ling & Joosten, 1999);
- Achieving excellence and best practice in teaching and learning (e.g. Dekkers & Andrews 2000);
- Introducing ICT into on-campus teaching (e.g. Dekkers & Andrews 2000).

Challenges that face universities for which ICT has been mooted as a possible solu-tion include:

- The risks for higher education in pursuing content-centred forms of curriculum delivery are growing as more and more employers are recognising the value to be gained from corporate institutions that focus on education and training based on the development of capabilities and competencies with demonstrated work-place application (e.g. Romeo et al. 2002);
- Flexible learning is still at the margins and likely to remain there for some time to come (e.g. Trevitt, 2000);
- ICT has been used to deliver courses and programs from Australia to cohorts in other countries. In Australia a large amount of delivery is undertaken in coun-tries in the South East Asia region. While the universities would consider them-selves to be implementing flexible delivery in their approaches, many of the students are finding the approaches inflexible and often contrary to their expec-tations and wishes. (e.g. Ziguras, 2001);
- Contemporary globalisation tends to put local traditions from many countries under pressure as globalised innovative practises are implemented (see Chapter 9 for a UK perspective on this issue).

By 2000 a number of universities within Australia had established purpose built campuses for the planning, development, support and evaluation of flexible delivery and flexible learning initiatives. For example, Logan Campus of Griffith University, The Ipswich campus of University of Queensland, Monash University and the Lilly-dale campus of Swinburne University. At this time most universities were implementing flexible delivery programs that were more about flexible teaching than flexible learning. The campuses built to support flexible delivery still used significant amounts of face-to-face teaching with students required to attend campus, few institutions using ICT were make flexible learning a reality.

By 2000, a significant number of universities within Australia had yet to have a stated policy on flexible delivery on their web sites. In a review of Australian universities, Dekkers and Andrews (2000) found that many institutions in 2000 were still in the process of articulating their definitions of flexible delivery and flexible learning and often used these terms interchangeably. The review also found that many universities had yet to formalise details of their management, organisation, or implementation. It was found that moves to explore and investigate flexible delivery and flexible learning were being undertaken by working parties working outside the normal functioning of existing bodies offering flexible courses (e.g. outside existing organisational units).

However, by December 2001 flexible delivery and flexible learning had derived momentum from the competitive environment now very evident in the higher education sector (Bell et al, 2002). The university sector had developed a significant interest in e-learning in that the move to an information and knowledge global economy had meant universities were now competing on a global scale. Over half the universities in Australia were offering fully online courses, with the majority (90%) at postgraduate level with the use of the Internet in course units ranging from 100% to as low as 9% in one university. These offerings were also being matched with online services such as library, off campus access, learning management systems, registration and enrolment, fee payment and support for students to support the flexibly delivered programs.

If university staff (academic and otherwise) are expected to relinquish their desks and telework at home this will also require a new style of management. The old-fashioned command and control system will no longer work. Managers will have to place more trust in their employees, because they will have less contact with them (Anon, 1998). They will have to find ways of motivating staff they cannot see. In teleworking it is important for staff still to feel part of the team, for example to be included in the organisation's social gatherings (Anon, 1998). Effective and successful staff recruitment will also become critical in the information age, as organisations cannot afford to lose their specialist workers (Anon, 1998). Santos, Heitor & Caraca (1998) suggest that the current departmental structure of universities needs to adapt into a more dynamic organisation with flexibility and become more informal. They suggest that these newer organic structures will be more innovative and easily adaptable to change.

ICT in Mainstream Teaching

Patterns of ICT uptake in mainstream teaching can be traced by reviewing the development of ASCILITE, the peak body in Australia promoting the use of ICT in learning. The home page of ASCILITE provides a brief history of the organisation and describes a number of small conferences held between 1983 and 1990 [3]. The early conferences were organised by a dedicated band of enthusiasts, not unexpectedly, from the computer science and education fields. The early conferences showcased innovative ideas using ICTs in classroom teaching and challenged other teachers to go forward with this information.

Even back in the 1980s the central tenet of such conferences was exploring ways to engage students and to provide student-centred learning settings through innovative use of the new technologies, which at that time were low power, locally networked desktop machines running a few pieces of application software. In 1985, for example, one of the keynote speakers was Diana Laurillard from the Open University in the UK who demonstrated and discussed the theme using videodisc and multimedia examples. The proceedings of the various conferences from the early days provide an interesting array of papers of both a scholarly and non-scholarly form (e.g. Isaacs, 1993) describing a plethora of teaching ideas and activities and using many different acronyms (e.g. CAI, CBL, CML and other such C commencing acronyms). Throughout the 1990s activity within the ASCILITE community grew and annual conferences were held across the country leading to larger and larger audiences and levels of active participation and interaction.

By 2000, a growing awareness of effective and meaningful teaching and learning was seeing useful synergies emerging between the use of ICT and the adoption of powerful learning strategies. Academics in Australia were realising that the Web was showing particular promise for supporting meaningful learning through its functionality, support for flexible delivery modes and capacity to link and connect those involved in the learning process (e.g. Duschatel & Spahn, 1996; Levin, 1999; Ferry, Hedberg & Harper, 1999). The possibilities now existed for rich learning based on this technology, but for the most part, pedagogically sound and exciting web courseware tools were yet to be developed to take advantage of such opportunities.

One of the key issues here was that the pace of change of emerging web technologies was so rapid that pedagogical models were not available to help create web tools from a learner-centred perspective (Bracewell et al., 1998). Salomon (1998) supported this concern that for the first time in history, technologies were outpacing pedagogical and psychological rationale. However, online learning practitioners were starting to get access to a body of literature that reported on innovative tools, with strong pedagogical underpinning. For example, Bonk & Cummings (1998) had reported on interactive tools for online portfolio feedback, profile commenting, and web link rating. Oliver & McLoughlin (1999) were building tools for online debate, reflection, concept mapping and student surveying and discussion. While Luca & Oliver (2002) were building tools to support peer assessment processes online.

Government Support

Since the 1990s, governments around the developed world have been attempting to withdraw from their responsibility of funding universities (Duguet, 1995). They are placing more emphasis on parental responsibility and encouraging universities to develop other sources of funding, for example by forming business partnerships. Governments are also encouraging the development of online courses because they realise how much cheaper it is to provide these courses. Studies have shown that the cost of providing distance education is half that of courses delivered in the traditional way, once initial costs have been covered (Duguet, 1995). Some of these figures are questionable but the divergence is clear. There are also many hidden costs in moving to e-learning, most of them to do with manpower issues and the issue of practicality.

In Australia, from the early days of ICT adoption in education and training, there were government initiatives aimed at supporting the early adopters of ICT as a strategy to increase the awareness of others and the levels of uptake into curriculum delivery. A number of government funded organisations were set up to encourage and support such activities among university academics.

Committee for the Advancement of University Teaching (CAUT)
The CAUT was established in 1992 to identify and promote good teaching, learning and assessment practice in higher education and to foster and facilitate innovation in higher education teaching. CAUT as an organisation followed from the Commonwealth Staff Development Fund (CSDF) established earlier. The visibility of CAUT among Australian academics was primarily due to the support it provided to individuals and universities to foster innovation and best practice in university teaching. For a number the years between 1993 and 1995, academics and institutions could bid to gain substantial amounts of funding to support the development and implementation of teaching innovations. CAUT funded nearly 150 year-long projects with amounts usually in the range of $50,000. The majority of projects involved applications of ICT in teaching and learning and a number of reports were undertaken at the time to assess the effectiveness of the project in supporting ICT uptake in Australian universities. The CAUT grants supported a wide range of innovative application of ICT in teaching and learning although a major report (Alexander, McKenzie & Geissinger, 1998) which investigated the effectiveness of the products reported that the majority of the ICT projects produced resources and intellectual property that remained within the participating institution with few projects producing materials that were used elsewhere.

Committee for University Teaching and Staff Development (CUTSD)
CUTSD was established to build on the work of CAUT and acted as its replacement. CUTSD was established for a three year period from 1997 to 1999 to promote quality and excellence in university teaching. The Committee made recommendations to the Minister for Employment, Education, Training and Youth Affairs for the funding of National Teaching Development Grants for projects supporting innovative teaching for individual academics and institutional organisational units as well as provid-

ing grants for academic and administrative staff development projects. Like its predecessor, it provided generous funding for university-based project and encouraged staff to explore opportunities for staff development within projects. Once again, the majority of funded projects looked to develop online and technology-based resources, but with broader contexts and applications than were evident in the CAUT projects.

Australian Universities Teaching Committee (AUTC)

The AUTC was established in 2000 to follow on from CUTSD, as part of the Government's commitment to promoting quality and excellence in university teaching and learning in Australia. The AUTC aimed to identify emerging issues in teaching and learning in Australian universities and administered a grants programme designed to identify and support effective methods of teaching and learning, promoting the dissemination and adoption of such methods across the higher education sector. While CAUT and CUTSD had provided funding for projects among individual academics and universities, the AUTC adopted a different strategic direction and supported more large-scale projects conducted by consortia seeking products with more general application and use. The majority of these projects provided information and resources, supporting technology usage in university teaching through more general activities such as teaching large classes, assessment practices and opportunities for flexible delivery of programs (see Chapter 4 for a UK perspective on nationally funded projects).

Education Network Australia (EdNA)

EdNA originated in 1994 with the intention of supporting educational institutions across all sectors accessing and using the Internet. EdNA has developed into a national framework for collaboration and cooperation and provides a number of services for schools, Vocational Educational and Training, and higher education (Mason, 1998). Among these are:

- EdNA Online, a web site servicing EdNA stakeholders and providing linkages between each of the sectors. Multiple pathways for information retrieval and resource discovery are an important feature;
- EdNA Online portal offering such services as access to quality, evaluated resources discoverable through the metadata repository, metadata exchange with other repositories, distributed searching of other, valued repositories, online collaboration via forums, chat, notice boards and discussion lists;
- EdNA metadata standards have been developed to support resource discovery and access in the local setting. Thee standards comprise a set of guiding principles together with a set of metadata elements and element qualifiers which are situated within the Dublin Core Metadata Initiative (DCMI) framework (Milea, 2003).

Reusability: an Admirable Attribute?

As part of the concern about efficiency in flexible delivery, two approaches to the use of multimedia content in online learning were starting to emerge in the late 90s, developing in parallel in Australia, and worldwide. The first was the development of generic learning designs based on our knowledge of best practice in terms of high quality learning. Significant advances were being made through projects such as the SoURCE project [4], which was developing instances of learning designs and an Australian University Teaching Committee (AUTC) project on use of ICT in learning (Harper et al, 2001) which was developing mechanisms to streamline this process through generic learning designs and symbolic languages. However, these, and similar projects were not addressing the reuse or management of resources in any specific way.

The second approach evolved from information systems and is characterised by the current research in learning objects. These projects, such as ARIADNE [5], MERLOT [6] and LRC [7] are based on developing standard descriptors [8] for media objects and then developing mechanisms to put these learning objects together as sessions or subjects in a learning sequence (Wiley, 2000). In Australia the federal government and all states, through the 'Curriculum Corporation', funded a national project on a K-12 (kindergarten to year 12) learning repository of learning objects, also useable in higher education in some contexts. The repository, being developed under the banner of 'The Learning Federation' [9] was to cater to the K-12 sector in all Australian states and Territories as well as New Zealand and deliver high quality learning objects, with simple sequencing tools for teachers to construct online learning environments. What is missing in these initiatives is a way of taking advantage of both of these approaches and developing a new way forward which would revolutionise the e-learning space i.e. a way of bringing together the power of quality learning designs and the concepts of reusability of learning resources as learning objects or digital items (see Chapter 3 for a fuller discussion of learning objects and reusability).

Alliances to Build Expertise and Offerings

As the drive to develop export income and cater for a broader range of students cost-effectively continued to drive the thinking about online learning in Australia in the late 90's, a series of alliances of various sorts started to be developed to optimise development costs and take advantage of the movement toward learning objects and repositories of learning resources. Early alliances within Australia included the Open Learning Agency, which was perhaps the first consortium of universities offering shared undergraduate courses and programs through technology-based flexible delivery, primarily for Australian students. This was followed by the Post-graduate and Graduate Education consortium (PAGE), which involved yet more Australian universities in technology-facilitated flexible delivery of graduate and post-graduate courses.

In recent years the consortia and alliances, which many Australian universities have joined (and helped to form) have moved to include overseas partners and international markets. For example, Universitas 21, an international network of 17 re-search-intensive universities in ten countries has developed large repositories of learning resources for use by its partner members. Its purpose is to facilitate collaboration and cooperation between the member universities and to create entrepreneurial opportunities for members on a scale that none could achieve operating independently or through traditional bilateral alliances. More recently we have seen the formation of the Global University Alliance, a partnership including a number of Australian universities and other fully accredited international universities. The Global University Alliance is "dedicated to providing students from around the world with accessible, rewarding educational experiences by leveraging the latest interactive web and data-based technologies, and making use of the expertise and resources of our global network of teachers and universities" [10].

A number of Australian Universities have also formed alliances with online brokers. For example, NextEd, a Hong Kong-based systems integrator, provides infrastructure, primarily in Asia, with local partners using the NextEd resources to deliver their courses. NextEd works with private and public education institutions, corporations, professional associations, and training organizations located in Hong Kong, China, Malaysia, Australia, Canada, Europe, New Zealand, the UK, and USA. NextEd has collaborated with a number of Australian universities to provide a capability for them to deliver courses and programs in countries throughout the region using this third-party technology-based infrastructure [11].

These activities demonstrate a trend within Australian universities to seek new markets for courses and to use the technology opportunistically in ways that add value. This is distinct perhaps from earlier activities, many of which were primarily aimed at improving the quality of learning in the sector. The alliances offer opportunities for universities to share resources and delivery systems. Within these activities, there is an underlying assumption that academics will readily share resources and describe them in ways, which will allow others to access them. There is an assumption that such resources can also be seamlessly integrated, by members, into their own courses and programs. These assumptions are yet largely untested and there is considerable effort and energy being expended worldwide to provide enabling strategies so that they might be realised.

Conclusions

The picture that emerges in the Australian post compulsory education scene, with respect to ICT, is one demonstrating significant growth, change and continual evolution of ideas and applications. As universities have developed their expertise with technology and as new technologies have emerged, universities have been very quick to explore new and innovative applications. In the current climate, many of the applications have an economic imperative and are serving well to assist many universities in achieving strategic their goals. There are still considerable issues, debate and conjecture around the maintenance of quality, consistency, personal and professional development as well as the requirement for often, significant infrastructure changes within our institutions.

Evidence from surveys in the 1990s suggest that around the world, universities appear to be slowly adapting to the use of ICT and are starting to change the structure of the organisation. For example, The International Islamic University in Malaysia started restructuring some years ago in order to contribute to Malaysia's development (Majid & Abazova, 1999). The internal political opinion was that ICT could help the country bring itself into the 'Information Age' and start gaining the skills that are enjoyed by more well- developed nations. Universities in the United States have joined the Internet 'frenzy' with relish, (Hall et al., 1999). Structural changes have radically altered both internal and external environments within higher education, affecting course changes, teaching delivery, administrative hierarchies, faculty and school boundaries. Such major upheaval is starting to bring about a new system of learning in higher education. (O'Donoghue, Singh & Dorwood, 2001)

Greenhill (1998) undertook a study of universities in Australia to examine how the structure of a 'virtual' organisation differs from an actual organisational structure. One of the main differences is based on the non-existence of a time factor in cyberspace. The only time factor used is the length of time one person is logged on and when they log off. Furthermore, information is transferred from one side of the world to the other so quickly it seems to take no time at all. Greenhill sees 'virtual' universities as being an extension of the real world due to this complete disregard for time as used in real life. To accommodate such measures a '24/7' curriculum is an extreme measure of re-structuring and most universities would probably be unlikely to take change quite this far and the time issue alone is seen as an insufficient reason.

It is important, to maintain the appeal of the current offerings by increasing study flexibility and pathways and using expanded access options to facilitate innovative educational delivery strategies. Pritchard and Jones (1996) identify the key enablers as continuous commitment to standardised content development, 'Intellectual Property' management, staff and flexible delivery support structures. Off campus exporting of education offers any university an opportunity to compete with a different teaching structure, a comparative advantage in terms of their ability to acquire teaching materials from perceived higher quality lectures and to package and broker these into their course offerings (Pritchard and Jones, 1996). The evidence suggests that if another paper of this form is written for the 20th anniversary of the Association for Learning Technology, we will see changes and developments of similar scale and magnitude as universities continue to use technology in clever and innovative ways to support all aspects of their business and enterprise.

References

Alexander, S., McKenzie, J., & Geissinger, H. (1998). *An Evaluation of Information Technology Projects for University Learning* [Video Tape]. Canberra: Australian Government Publishing Service.

Anon (1998). Teleworking Opens up New Possibilities. *Decision (IRL)*, 2, 6, 20-23.

Baumbach, D., & Bird, M. (1998). *The 7 Habits of Highly Effective TechnologyUsing Educators*. Florida USA: FETConnections, FETC.

Bell, M., Bush, D., Nicholson, P., O'Brian, D., & Tran, T. (2002). *Universities Online: A survey of online education and services in Australia*. Commonwealth Department of Education, Science and Training. Occasional Paper Series. Higher Education Group.

Bonk, C. J., & Cummings, J. A. (1998). A dozen recommendations for placing the student at the centre of web-based instruction. *Educational Media International*, 35,2, 82-89.

Bracewell, R., Breuleux, A., Laferrière, T., Benoit, J., & Abdous, M. (1998). *The emerging contribution of online resources and tools to classroom learning and teaching* (Report submitted to SchoolNet/Rescol by TeleLearning Network Inc.) [On-line]. Available: http://www.tact.fse.ulaval.ca/ang/html/review98.html

Catchpole, M. (1993). Interactive media: the bridge between distance and classroom education. In T. Nunan (Ed.), *Distance Education Futures*. Adelaide: University of South Australia.

Catchpole, M. J. (1992). Classroom, open and distance teaching: A faculty view. *The American Journal of Distance Education*, 6, 3, 34-44.

Clark, R. E. (1983). Reconsidering research on learning from media. *Review of Educational Research*, 53,4, 445-459.

Clark, R. E., & Salomon, G. (1986). Media in Teaching. In M. C. Wittrock (Ed.) *Handbook of Research on Teaching*. 3rd Ed (pp.464-478). New York: Macmillan Publishing Co.

Coaldrake, P., & Steadman, L. (1998). *On the brink: Australia's universities confronting their future*. St Lucia: Queensland University Press.

Cuban, L. (1986). *Teachers and machines: The classroom use of technology since 1920*. New York: Teachers College Press.

Dekkers, J., & Andrews, T. (2000). *A meta-analysis of flexible delivery in selected Australian tertiary institutions: How flexible is flexible delivery?* Paper presented at the ASET HERDSA, Toowoomba.

Duguet, P. (1995). Education: Face – to- Face? *OECD Observer*, 194,17 – 21.

Dunnett, C. (1990). Open access- open media. *Educational Media International*, 27,4,198-207.

Duschatel, P., & Spahn, S. (1996). *Design for Web-based learning*. Paper presented at the WebNet'96, San Francisco, USA.

Evans, T., & Nation, D. E. (1993). Distance education, educational technology and open learning: Converging futures and closer integration with conventional education. In T. Nunan (Ed.), Distance Education Futures. Adelaide: University of South Australia.

Ferry, B., Hedberg, J., & Harper, B. (1999). Designing computer-based cognitive tools to assist learners to interpret graphs and tables, *Australian Journal of Educational Technology* 15, 1, 1-19.

Garrison, D. R. (1985). Three generations of technological innovations in distance education. *Distance Education*, 6,2 , 235-241.

Greenhill, A. (1998). Commodifying *Virtual Education: Virtual Classrooms, Universities and Virtual Organisational Existence*. Paper presented to "Commodification" Conference, Wollongong, Australia.

Hall, R., Butler, L, Kestner, N., & Limbach, P. (1999). Combining Feedback and Assessment. *Campus Wide Information Systems*, 16, 1, 24.

Harper, B., O'Donoghue, J., Oliver, R., & Lockyer, L.,(2001) New Designs for Web Based Learning Environments, In C. Montgomerie & Viteli, J (Eds.) *Proceedings of ED-MEDIA, World Conference on Educational Multimedia, Hypermedia and Telecommunications* (pp 674-5).

Hedberg, J. G., & Harper, B. M. (1996). Supporting and developing teachers through telecommunications *Educational Media International*, 33,4, 185-189

Hutchison, C. (1998). The Virtual University and the Culture of Learning. *Education Libraries Journal*, 41,1, 5-11.

Inglis, A., Ling, P., & Joosten, V. (1999). *Delivering digitally: Managing the transition to the knowledge media*. London: Kogan Page.

Isaacs, G. (1993). So What's New? ASCILITE '83 to ASCILITE '93. *In Proceedings of ASCILITE 93*. [On-Line]. Available: http://www.ascilite.org.au/history/ASC93isaacs.html

Kelly, M. (1987). Barriers to convergence in Australian higher education. In P. Smith and M. Kelly (Eds.), *Distance Education and the Mainstream* (pp. 175-200). Beckenham: Croom Helm.

Kozma, R. (1991). Learning with media. *Review of Educational Research*, 61,2, 179-211.

Levin, J. (1999). Dimensions of network-based learning. *International Journal of Educational Technology*, 2,10, 12-21.

Luca, J., & Oliver, R. (2002). Developing an instructional design strategy to support generic skills development. In A. Williams, C. Gunn, A. Young & T. Clear (Eds.), *Proceedings of the 19th Annual Conference of ASCILITE*. Auckland, NZ: UNITEC University of Auckland, (pp 401-412).

Majid, S., & Abazova, A. (1999). Computer Literacy. *Asian Libraries*. 8, 4, 100-109.

Mason, J. (1998). *EdNA - a Historical Snapshot*. Paper presented at the University of Melbourne Faculty of Education Graduate Student Conference, December 1998. [On-line]. Available: http://www.edfac.unimelb.edu.au/insight/postscriptfiles/vol1/mason.pdf

Milea, J. (2003). *The EdNA Metadata Standard*. Paper presented at the DC-ANZ Metadata Conference, February, 2003. [On-Line]. Available: http://www.educationau.edu.au/papers/edna_metadata.pdf

Moore, M. G. (1989). Three types of interaction. In M. G. Moore & G. C. Clark (Eds.) *Readings in Principles of Distance Education*. University Park, PA: American Centre for the Study of Distance Education.

Nipper, S. (1989). Third generation distance learning and computer conferencing. In R. Mason & A. Kaye (Eds.), *Mindweave: Communication, Computers and Distance Education*. Oxford: Pergamon.

Nunan, T. (1996). *Flexible delivery – what is it and why is it part of the current educational debate.* Paper presented at the Higher Educational Research and Development Society of Australiasia Annual Conference, Perth, July 1996.

O'Donoghue, J., Singh, G., & Dorward, L. (2001). Virtual Education in Universities: A Technological Imperative. *British Journal of Educational Technology*, 32, 5, 517-530.

Oliver, R., & McLoughlin, C. (1997). Interaction patterns in teaching and learning with Live Interactive Television. *Journal of Educational Media*, 23,1, 7-24.

Oliver, R.,& C. McLoughlin (1999). Curriculum and learning-resource issues arising from the use of Web-based course support systems. *International Journal of Educational Telecommunication* 5, 4, 419-438.

Pritchard, T., & Jones, D. (1996).*Open Learning And/As The Virtual University, University Of Melbourne, The Virtual University?* Symposium, Melbourne, Australia, November 21-22.

Reeves, T. (1992). Evaluating schools infused with technology. *Education and Urban Society*, 24, 4, 519-534.

Reeves, T. (1993). Pseudoscience in Computer-Based Instruction: The case of learner control research. *Journal of Computer Based Instruction*, 20,2, 39-46.

Romeo, G., Webster, L., Varsavsky, C., & Edwards, S. (2002). *Interlearn – Linking learners at Monash University*. Proceedings of Australian Council Educational Computer Conference, Hobart, July 2002.[On-Line]. Available: http://www.tasite.tas.edu.au/acec2002

Ross, B. (1993). What has happened to convergence? In T. Nunan (Ed.), *Distance Education Futures*. Adelaide: University of South Australia.

Salomon, G. (1998). Novel constructivist learning environments and novel technologies: Some issues to be concerned with. *Research Dialogue in Learning and Instruction*, 1, 3-12.

Santos, S, Heitor, M., & Caraca, J. (1998). Organisational Challenges for the University. *Higher Education Management*, 10, 3, 87-107.

Schramm, W. (1977). *Big Media, Little Media*. Beverley Hills, CA: Sage Publications.

Singh, G., O'Donoghue, J., & Betts, C. (2002). A UK Study into the Potential Effects of Virtual Education: Does On-line Learning Spell an End for On-campus Learning? *Journal of Behaviour and Information Technology*. 21, 3, 223-229.

Taylor, J. C. (1992). Distance education and technology in Australia: A conceptual framework. *Council of Distance Education Bulletin*, 8, 22-30.

Trevitt, C. (2000). *Flexible learning in higher education: Examining the case for the learning profession and the learning discipline*. Proceedings of 6th ALIA Conference, Canberra, 2000.

Wiley, D. A. (2000). *Learning object design and sequencing theory*. Unpublished
    Doctoral Dissertation, Brigham Young University, Provo, UT. [On-Line]. Avail-
    able: http://davidwiley.com/papers/dissertation/dissertation.pdf
Ziguras, C. (2001). Educational technology in transnational higher education in
    South East Asia: The cultural politics of flexible learning. *Educational Technol-
    ogy and Society*, 4,4, 8-18.

Notes

[1] Open Learning Agency: http://www.ola.edu.au
[2] Flexible Learning and Higher Education Resources:
http://bridge.anglia.ac.uk/www/flexi.html
[3] ASCILITE: http://www.ascilite.org.au/history.html
[4] SOURCE: http://www.source.ac.uk/
[5] Ariadne: http://www.ecotec.com/sharedtetriss/projects/files/ariadne.html
[6] MERLOT: http://taste.merlot.org/
[7] LRC: http://www.edtec.unsw.edu.au/frames_inter.cfm?are_a=2&page=S_LRC.cfm
[8] LTSC, Learning technology standards committee web site http://ltsc.ieee.org/
[9] The Learning Federation: http://www.thelearningfederation.edu.au/tlf/
[10] Global University Alliance: http://www.gua.com
[11] NextEd: http://www.nexted.com

9

Facing The World Challenge: Risk, Resistance and Collaboration

Gabriel Jacobs

In 1997 the Dearing Report presented its vision of higher education in the decade then to come. It was to be a period when there would emerge a flourishing world wide e-learning market set to dominate post-compulsory education (Dearing, 1997, Section 13.8). Similar predictions followed, many gurus confidently expecting a fabulous treasure chest very soon to be opened (see Keegan, 2000). The Dearing Committee contended that if higher education in the UK was not prepared to take advantage of the potential bonanza this forthcoming development represented, it risked disastrously losing out to international competition. Besides, so stated the Report, whether the opportunity was lost or exploited, for the majority of UK students the delivery of course materials would within the decade be computer mediated (Section 13.3). In my response to the Report (Jacobs, 1997), I maintained that the Association for Learning Technology (ALT) had urgently to decide how to capitalise on its forecasts, the most critical area on which to focus being, in my view, one in which I believed that ALT had seen less success than had been hoped for. This was in ALT's aim to ensure the widespread acceptance by senior management within higher education of the need to change tack in the direction of educational technology, which would allow the UK to muster collaborative forces against the impending onslaught from overseas. Dearing, I argued, gave ALT the authority with which to further that aim. In this chapter I will review the extent to which post compulsory education institutions in the UK have responded to the challenge of developing a world wide e-learning market and discuss factors that are influencing this response.

Global-Wise, Where are we?

With only four years to go before the end of the post-Dearing decade, it has to be said that we are far from reaching the point at which we can sit back and be satisfied; a situation which has recently been confirmed in the Government White Paper (DFES 2003) on the future of UK higher education, where the current lack of collaboration between institutions is highlighted as a specific failing (Section 1.19). The competitive market in which UK post compulsory education finds itself remains stubbornly localised, intense rivalry between institutions for student recruitment being far more the order of the day than a collective defence against global competition. This is partly, perhaps principally, because of doubts about how well stocked the supposed treasure chest actually is. Computers are certainly used more for learning and teaching in today's universities and further education colleges than they were in 1997, but not to the extent expected by Dearing, and even less as global distance-learning tools. A glance at the programmes and proceedings of conferences concerned with post compulsory education published over the last few years reveals an inexorable rise in the number of contributions related to the use of the Internet as a teaching and learning facility, either as an information resource, or used interactively in conjunction with a physical classroom, or – less commonly – to replace a physical presence entirely: ALT conference programmes are prime examples (Jacobs, 2001). Nor can it be doubted that the Internet offers unparalleled mass educational opportunities of which no previous age could hardly have conceived. So it is not surprising if, since even well before Dearing, its obvious advantages have regularly been acknowledged and described in countless books, articles, papers, reports, political pronouncements, and everywhere in the medium itself. In spite of this, pick-and-mix predictions of the 1990s, in which vast cohorts of UK students would be accumulating transferable credits by remotely studying a flexible portfolio of modules at different institutions, even Europe-wide and beyond, have still not materialised (and this in the face of certain powerful European Union initiatives in that direction).

Not withstanding the advent of tuition fees and the effective abolition of maintenance grants, with the consequent need for many so-called full-time students to work half-time or more behind the counters of McDonald's, the UK higher-education student culture is still one in which the social life essentially unavailable in virtual environments is considered to be as important as academic study, if not more so. Hence the high profile of sports facilities, leisure activities and prospective social interaction to be found on the web sites of many institutions, in their printed prospectuses and promotional brochures, and in their advertising in the media. There are countries where distance learning is progressing at a somewhat faster rate than in the UK, but they are not necessarily similarly placed, given, for instance, that geographical location involving large distances and/or extremes of weather often play their part in encouraging online services (see Chapter 8) Yet even in locations where long outreach would seem to be the obvious path to increased student recruitment, considerable resistance to using communications technologies to achieve it is still to be found. Thus in parts of the USA, where computer-mediated distance education has so far seen its most dynamic implementation, academics working in universities with potential catchment areas the size of a small nation may still face apparently insurmountable difficulties in overcoming management opposition when it comes to online course delivery (Beckstrand, 2002). Education is a reflection in this respect of industry and commerce, where the clearly identifiable gains of teleworking are often rejected by companies whose work culture is too firmly entrenched for risks to be taken with change. Barring special circumstances, at the moment the natural choice of young people, is to 'go to' college, as workers 'go to' work.

Balancing the Risks

Many who are involved in higher and further education nevertheless feel that Dearing's vision of the future is destined eventually to become reality. This is unmistakably to be perceived in, for example, the late David Squires often cited (though self-admittedly speculative) notion of 'Peripatetic Electronic Teachers' offering educational services to learners on demand whatever their location, in this way breaking the long-standing link between the teacher and the single employing institution (Squires, 1999). But if a decisive change of direction is to happen, it will happen above all for self-interested reasons. Let us leave to one side what have become the platitudes, however accurate or otherwise, surrounding the promise of new technology for students' learning: the infinite patience of computers and their natural attraction for the young (still continually cited even in our post-Skinnerian educational world), online access to the planet's finest expertise, the flexibility to work at one's own pace and in one's own time, the potential for the liberation of identity ('I am who I am wherever I am') and so forth. If the past is any predictor of the future, such factors may well be acknowledged by administrators, but in most instances the acceptance of their benefits as express advantages for students will not in themselves result in generous funds being allocated to the use of digital technology for remote learning, even when costs and risks can be shared in collaborative projects. In other words, if the predictions of Dearing are to come about, they will not do so as a result of the altruistic intentions of senior management, even though such intentions may be praiseworthy. No, the main institutional force for a change of direction will not be pedagogical, but hard-boiled educational politics entrenched in matters of finance. The attitudes of administrators towards learning technology have unquestionably changed over the last ten years, but the change has been slower than might have been hoped for, at least partially because it has come about under steady governmental pressure (itself the result of public pressure) which translates into funding, rather than through a desire, however sincere, to accomplish a worthy mission. In his report on the uptake of new technologies within higher education, Smith (2001:1) identified the major barriers to online provision of services as "the academic staff time and resources required to develop materials, fear of technology amongst academic staff, the emphasis on research in many institutions, and the lack of institutional learning technology strategies". All these factors are real enough, but apprehension among administrators faced with what they perceive as risky investment has surely always also been at the core of their inertia, and remains so.

Added to this is an impression of senior managers paying lip service to the Internet but without a deep commitment to seeing it used to its full potential as a teaching and learning medium able to reach beyond borders. I confess that I can claim for this only the anecdotal evidence of the comments of many colleagues and my own experience. For example, I have more than once, and quite recently, heard certain senior managers talk of the importance of new technologies while taking a kind of perverse pride in pronouncing, with a smile, their comparative ignorance of them ("My ten-year-old knows far more about computers than I do"). It is perhaps natural, then, that such administrators may be willing only to back a few field trials carried out by eager young staff, and then perhaps only when such trials are supported by a fixed-term external grant, or to join forces with other institutions only when the initial investment is low, or better, when it can be delayed until there are explicit signs of a guaranteed return.

This is unsurprising, too, when one takes into account stories of failed initiatives making the educational headlines, such as news that some US universities are pulling out of distance-learning programmes because they have been shown to be making sizeable losses when costs are looked at with a cool eye, and hidden subsidies rigorously discounted. For example, the much heralded Fathom programme and associated web site [1] of New York University, Temple University, and Columbia University is now no more than the embers of a once claimed blazing educational fire fanned by the prestige of Cambridge University Press, the Smithsonian Institution, the Universities of Chicago, Washington and Michigan, and the LSE. The sense of disillusionment lying behind the artificially up-beat message on the home page of that web site is plain enough; essentially, that in the face of outstanding success the initiative has been abandoned! Further, while major US players in the e-university market, the likes of Harvard Business School and Stanford University, still proclaim full commitment to their online ventures, not one has yet been able to fulfil the initially trumpeted high hopes of visibly profitable computer-mediated distance courses. They have come to be seen, maybe even see themselves, as survivors rather than trail-blazers. Initiated before the 'dot-com' bubble imploded, therefore at a time when e-commerce appeared to be the only real future for business, they have discovered that the delivery of online courses is no easy sell. As in so many commercial areas, the market has been found to be smaller and less enthusiastic than anticipated. This is likely to be because the self-discipline, strong motivation, and high degree of control over one's own learning required to complete an online course is in shorter supply than many visionaries assumed. It is well known that Cardean University, UNext's qualification-granting institution supported by names as respectable as the LSE and Carnegie Mellon University, has seen far fewer students than expected pass through its virtual doors since they opened over three years ago. This is in spite of a huge initial financial investment and a widely held belief at the start of the project, even in official US circles, that each participating institution could expect to be receiving around $20 million a year in royalties within five to eight years (CHEA, 2000).

Concomitant with wildly optimistic estimates of future income is the lack of dependable information with respect to expenditure other than the sketchy details, which become available when, as noted above, institutions decide to retire from the online arena once having meticulously calculated costs. There is no equivalent of, for example, the TCO (Total Cost of Ownership) concept for companies' investment in information systems. Condron and Sutherland (2002:8) note, with justification, that the gap in the provision of expert advice on the actual costs of online provision has become a serious hindrance to institutional commitment. Not to mention stories that from time to time hit the media involving squabbles, which can only sully the image of online course delivery. For example, the decision this year by the UK government to attempt to reclaim a large European Social Fund grant made to the virtual university Virt-u.com (which collapsed after bitter arguments and then became a pornographic web site) hardly enhances the reputation of online courses. Nor do spam emails offering university degrees online, from BSc to PhD, with no entrance tests, interviews or assessment, and confidentiality assured. No wonder that several prominent organisations in the USA, such as the American Bar Association, simply refuse to recognise online-only degrees.

Yet it is not just a matter of financial or reputational risk for those in charge of higher education institutions: communications technologies seem also to pose an insidious threat to traditional academic culture; not a direct threat to teaching jobs, instructional technology does not lead to teacher-redundancy, regardless of the views of a few extremists who periodically proclaim the demise of the lecturer in favour of the machine (see Brabazon, 2002: 111), but to the age-old moral mission of academia. The powers-that-be, at least in traditional universities, often resist moves which appear to threaten that mission, and it can be argued that they are right to do so when it comes to the serious competition represented by the private sector. Seen as particularly menacing is the intrusion of large corporations with their profit motives, which stand in contrast to what ought to be the goal of higher education, namely the edification of citizens and thus the 'Public Good'. The perceived danger with for-profit initiatives, which seek to benefit from the status of universities is that their penetration of the higher education environment can impinge on academic autonomy. When, for example, the giant 'Thomson Learning' works in close conjunction with university partnerships, something that makes commercial sense for all sides, its understandable intention is to see its own distance learning resources used, and even protected against 'contamination' by individual academics. It can defend such an intention on the perfectly reasonable grounds that educational partnerships will not succeed unless they accept common course structures and materials, but the freedom of individual academic institutions is nonetheless potentially compromised. Or again, universities are naturally tempted by offers from well-heeled companies to equip computer laboratories in return for installing proprietary products, and even for offering proprietary qualifications. Administrators, perceiving the risks, should perhaps indeed be wary of developments of these kinds. Harmonization is plainly an important ingredient for the success of collaborative educational efforts, but it also means an inevitable rigidity, while private-sector money can result in a damaging tension between commercial objectives (typically not extending much beyond a year or two, and even to as little as a financial quarter) and the longer-term ideals to which all educational institutions should surely be subscribing. As powerful private interests, attracted by the image of a seemingly lucrative online education market, seek to use university standing to validate courses, senior managers, above all in research universities, are often torn between, on the one hand, the lure of private-sector money with its strings attached and its utilitarian, market-driven targets, and on the other the desire to maintain the time-honoured sovereignty and independence of the Academe. Tiffin and Rajasingham (1995:165), in their book on the virtual classroom, published a good while before the private-sector rush into online education had begun in earnest, identified the potential dilemma with admirable foresight. Their words are worth quoting since they resonate today even more powerfully than eight years ago:

> If educational purposes are too tightly linked to [...] market forces they will become narrowly utilitarian and countries will lose the goals of education to develop citizens with social and cultural skills as well as work skills.

Many senior managers who came to their positions of authority before the advent of an extensive private-sector desire to join forces with the public sector, that means most senior managers, would probably concur.

What is more, senior managers are hardly to be tempted down online avenues by persistent reports of student pressure for more face-to-face teaching. It might be thought that online delivery, especially given its potential for interactivity, considered by Phipps and Merisotis (2000:16) to be 'the *sine qua non* for quality in distance learning', would be exceptionally attractive to the new, supposedly computer-literate, generation of students. But, 'it ain't necessarily so', as a number of surveys have found. My own experience, and that of many of my colleagues, is that what students want, if anything, is the very passivity that conventional face-to-face lecture courses, complete with handouts and/or notes posted on an intranet, have long provided. Fundamentally, this is the same appealing passivity offered by non-interactive broadcasting, film, newspapers and magazines. The point is demonstrated by this year's case at the University of Bath in which students complained that contact-time with teachers had been reduced to unacceptable levels with the increase in web-based course delivery (Baty, 2003), or by Sussex University's radical restructuring of its Arts degrees to incorporate a much higher level of real contact between students and tutors (Woodfield, 2003). The result is that computer-mediated education is most often seen by institutional decision-makers as an add-on, necessary in the light of the hyperbole surrounding it, but not a core activity.

Reactive Drivers for Change

The perceived risks are thus manifold, yet significant change may well be on the horizon, if for no other reason than that it will be difficult for administrators and teachers alike, whatever their views, to stem the tide of technology pulled inexorably along as it is by user-expectations. Academics are hardly thought of with warmth and generosity by government officials, nor by the public at large. Especially as regards the so-called upper-echelon universities, they are seen by many as arrogant, resistant to change, living at the top of their ivory towers with their heads in the clouds. That view may be undergoing something of a transformation as young Turks with bright, so-called relevant ideas enter the profession, and as the State calls all academics to account with quality-assurance exercises, which attempt to measure their abilities and effectiveness as teachers. But universities are big ships to turn around. Such perceptions of academics make it easy to understand favourable government reactions to the idea of remote teaching using the latest technologies. If the Internet can be used to deliver courses devised and administered by selected forward-looking teachers, the remaining fuddy-duddy academics can be cut down to size, or if they are top researchers, sequestered in their offices and laboratories located in a dozen or so centres of excellence. Here is one impetus for change, which if one can be allowed to read a little between the lines of the White Paper (DFES 2003), appears to be backed by the government. When the Secretary of State for Education and Skills insists that we should "be thinking hard about whether institutions could do more to help the best researchers focus on research, rather than teaching [...]" (Section 1.17), one is tempted to see the writing on the wall (and maybe to wince at the apparent abandonment of the notion of research informing teaching and vice versa).

A further potential driver for change, which again has to be set firmly against the various qualms and reservations among senior administrators, is the fear of being left behind, a fear which is steadily increasing as competitor institutions announce moves, whether or not they are really significant, towards more online delivery, and as instances of promised or even actual success appear to signify the coming of a radically modified educational world. Examples of success in the USA would include the University of Phoenix Online [2] (widely acknowledged as the most thriving American virtual university), the University of Maryland's substantial distance-learning programme [3] and Universitas Global [4] (an amalgam of several American business schools and a private firm targeting Asia and South America). The American Futures Project web site [5], dedicated to promoting its vision of higher education in the USA and regarded as a reliable source of information, has continued for a long time to claim that the market is seeing an unstoppable globalisation of the sector, while 'the rapid advance of new technologies, and shifting demographics, are driving the system toward change'. In Australia, there is significant progress towards online provision (see Boezerooy, 2003) even though some of its advocates complain that the movement is frustratingly slower than in the USA (see, for instance, Cunningham, 2000). And in the area of the international pooling of resources we are witnessing the advent of initiatives involving really big names covering a wide range of subject-areas, such as the alliance (AllLearn) of the Universities of Yale, Stanford and Oxford [5]. For the moment, AllLearn offers only non-credit-bearing courses, such that, as Loncraine (2002) has it, the venture "sounds a bit like an Oxford tutorial, except that you provide your own sherry and there are no essays", but exploiting the thirst for learning for its own sake could yet prove a good way of establishing a solid online presence before moving into the world of cyber-graduation ceremonies.

Some UK Initiatives

In the UK, the premier example of innovative performance in distance learning is, of course, the Open University, although not until lifelong learning shot up the agenda, with the mature student market finally seeming to traditional institutions to represent some untapped reserves, was the Open University generally held by them to be a major threat. Its failure to penetrate the US educational scene after a multi-million dollar investment has done something to confirm the perception by the traditional institutions of the uncertainty of the international distance-learning market, but its increasing use of educational technology to deliver undergraduate courses has also made policy-makers in conventional universities watch its development very closely, later (if and when) perhaps to be pulled along by it.

Other, smaller initiatives are to be found in almost all UK universities and colleges. They are all to be commended, although one must always take success stories with a level of circumspection. It is to be expected that those involved with learning technology will not only display a certain bias towards the field, but will also seek ways of promoting the individual projects with which they are concerned. I was flooded with replies in March of this year when I put out an email call among ALT members for case studies of computer-mediated distance programmes, which might be suitable for use in this chapter. While it is gratifying to have received such an overwhelming response, indicating an admirable, and much appreciated desire to help a colleague, it is also clear that many who replied saw, understandably, an opportunity to raise the profile of their schemes. Many offered forecasts of future student enrolment, and some gave hints of profitability, even if candidly moderated with "allegedly" or some similar observation. Very few gave any indication of the here-and-now evidence of profitability and sustainability I was really seeking. And the majority involved either English-language teaching programmes (where there is indisputably a volatile, highly competitive and clearly lucrative market), or relatively small postgraduate programmes, many of the latter tapping the mature student (especially the MBA) market. There was a distinct paucity of responses claiming success, collaborative or otherwise, in undergraduate distance-learning programmes.

It has to be admitted that at all educational levels the UK picture is patchy and anecdotal, not to say chaotic, and the available information in any case unreliable in view of the fact that the definition of an online course is fuzzy: it can indicate anything from a simple intranet threaded-message facility to a fully fledged interactive module delivered at home and overseas with hardly any human intervention (though the latter is a comparatively rare animal). What is required in order to make a proper assessment of online delivery in the UK, although it would be a major undertaking fraught with all manner of obstacles, is a register of online courses, with full statistics on the range and level of learning-technology usage, independent estimates of true financial cost, and so forth. Without such data, the picture is bound to remain indistinct, something that can only add to the doubts felt by administrators within post compulsory education. But the wind of change, in part out of the control of such administrators, may well prove uncontainable: as the song (more or less) has it, 'when an old immoveable object meets an irresistible force, something's got to give'. It is unlikely that what will give will be customer demand.

A Rose-Pink Future?

Thus, despite all the fears and uncertainties, it may be that we are for once truly on the verge of a sea change, and for reasons not only related both to market forces but also, somewhat ironically, to the very risk aversion of post compulsory education administrators. The alternative, or at least a complement to Squires' model of the electronic teacher decoupled from a single institution is the one adopted by UK eUniversities Worldwide, a promising scheme despite certain vehement attacks on it as a shameful waste of public funds (see, for example, Greenalgh, 2003). By signing up to eUniversities, institutions can run small externally funded programmes with, consequently, only minimal risk (up-front costs are covered, although courses have to pay their way over time). They can also be seen to be joining the online band-wagon while remaining more or less independent and having some of the burden-some aspects of course administration reduced. The initiative thus appears to be just what is needed for the development of a genuine UK coalition able fend off interna-tional competition. There are, though, two potential impediments to the project's prospects as a winning response to that competition. The first is that while it offers a measure of course standardization, as yet it lacks a clear brand image for quality assurance. True, this is to some extent offset by the fact that in spite of being a commercial company, it is an integral part of the whole public higher-education sec-tor, in this way differentiated from individual consortia of universities. It remains to be seen, however, if the quality assurance it offers will have the standing its initia-tors hope for. The second is that it is principally aimed at the postgraduate market, leaving the undergraduate counterpart still without a similar seemingly viable model. In a personal communication to me, Jonathan Darby, Chief Architect of the project (and former President of ALT) defends the decision to concentrate almost exclusively on postgraduates by insisting that "e-learning really comes into its own with remote adult learners where the flexibility and adaptability it offers can be the mechanism that enables study alongside other work and personal commitments". If this is a valid view, it is obviously the right line for eUniversities to take in ensuring its own survival, but this does then leave a gap in the wider picture. Shortly after UK eUniversities was launched, Smith (2001, Table 1) saw this potential gap as a cause for concern, stating that HEFCE should ensure "fulfilment of the aims and objec-tives for which the project was originally set up", in particular "that the social inclu-sion agenda remains a priority, primarily through the development of undergraduate courses". It should be noted in this context, however, that the Open University is evolving towards greater reliance on e-learning in its core undergraduate business, even if that business is currently aimed principally at the lifelong-learning market as opposed to the immediate post-secondary market of 18 to 23 year-old full-time un-dergraduates. The long-established London University External degree schemes[7] are also increasingly turning to the use of Virtual Learning Environments for course delivery with the appearance of new online undergraduate degrees, although it is too early to assess their future success and sustainability. In any case, despite the com-paratively high profile of various initiatives involving research universities, the post-1992 institutions in reality tend to be more active and more forward-thinking in the area of local and global online services (not unexpectedly, the majority of the respondents to my call for case-studies were working in post-1992 universities). One can perhaps accordingly expect them to be taking the lead in future online undergraduate provision.

Whatever the case may be, and whoever is to lead the way, it is clear that young people around the world are sensitive to the potential of the Internet. It is becoming commonplace, but no less valid, to say that students are seeing themselves increasingly as customers, with the right to choose (and to complain). The future of education is seen by them and their parents as one offering consumer choice which, given the ubiquity of the Internet, could indeed translate into a proliferation of courses delivered from anywhere, and places of post compulsory education fiercely competing with each other for distance-learning students who pick their modules from any number of institutions, just as they vie now for their physical presence. A further small but not all that insignificant indication of this is that most virtual learning-tools are or have become web-based. The fact that such tools have been constructed and marketed in the way they are, in itself suggests a belief by hard-nosed commercial developers that education is about to undergo an Internet revolution, which by a kind of 'Parkinson's law' for learning technology will ultimately involve worldwide reach.

Conclusion

It has been my intention in this chapter to temper zeal with a pinch of pragmatism, possibly a dubious stance to adopt in a book which (justifiably) celebrates a decade of achievement, although am I not alone in distancing myself somewhat from the overstatements of the apostles of technoculture: take, for instance, the tenor of some of the excellent essays in Robins and Webster (2003). I began by talking of the predictions of Dearing and others of a rapidly emerging online post compulsory education market and the urgent need for UK institutions to exploit it. Similar predictions and warnings continue to be published. Molyneux (2003), for example, while doubting that UK higher-education institutions are ready to compete world-wide because they 'lack vision of leadership as we move into the "knowledge" economy', adds that 'if we are to compete in the global education services arena, we must move with the times.' How many times, and for how long, have we heard that? The problem for senior managers has always been that of being sure that the times, and circumstances, are right. Can they even be certain, for instance, that the funding authorities see online course delivery as a really urgent priority and will thus give it the full financial backing it deserves, whatever the intentions of various funded organizations which seek to promote it, whatever the seeming import of Government backing for eUniversities? O'Leary (2003) talks of the Government's unambiguous commitment to it, evidenced, he claims, by the White Paper (DFES 2003) which he sees as treating e-learning as "a natural part of the scene, rather than parcelling it off into a separate section", the proof being that "dotted through its 100 pages was a recognition of the growing role played by new technology". In fact, the use of learning technology is the subject of only four very brief mentions and is nowhere linked with global reach except tangentially as a brief reference to the fact that about 12 per cent of Open University students are studying from abroad. (Section 5.25). Can we nevertheless take heart from the clear statement, in the same section, that "HEFCE will now work with partners on plans to embed e-learning in a full and sustainable way within the next ten years"? It is to be noted, at least, that this timescale adds a further decade to the six years, which have already passed since Dearing. It was John Kenneth Galbraith who said that there are two classes of forecaster: those who don't know and those who don't know they don't know. We should be wary of the latter, and accept only with a dollop of suspicion self-assured predictions that a global educational revolution based on learning technologies is just around the corner and will therefore automatically attract huge 'defence' spending.

On the other hand, to ignore the clear possibilities of learning technologies is surely the real risk to which many of those involved in one way or another with them regularly call attention. If the last few years are anything to go by, the movement will continue to be a creeping one, but critical mass may finally be within sight. The last decade has been one of constant flux in the area of the educational use of information and communications technologies (a fact reflected in the shifting fashions for its acronyms: IT, ILT, CIT, C&IT, I&CT, ICT, and so on), so much so that in the time between my writing this chapter (May 2003) and the few months before the book appears in print, the landscape will almost certainly change, perhaps dramatically, making some of what I have had to say outmoded. Yet for the longer term, as, say, ALT celebrates the first 20 years of its existence, there is real reason to suppose that its members will no longer find themselves battling against resistance or inaction but rather, the war essentially having been won, spending most of their energy in supporting the fine-tuning of a universally accepted, widespread, collaborative online provision to all categories of student both in this country and abroad. The future may be uncertain, but it cannot be called unpromising.

References

Baty, P. (2003). Bath gets nod for double vision. *THES*, 7 March.

Beckstrand, S. (2002). *A Comparison of Distance Education Delivery Methods in Southern Nevada*, Ph.D. Thesis, University of Teesside.

Boezerooy, P. (2003). *Keeping Up with our Neighbours*. York: LTSN-GC.

Brabazon, T. (2002). *Digital Hemlock*. Sydney: University of New South Wales Press.

CHEA (2000). *Distance Learning in Higher Education*, Washington DC: Council for Higher Education Accreditation. [On-Line]. Available:
www.chea.org/Research/distance-learning/distance-learning-3.cfm

Condron. F., & Sutherland, S. (2002). *Learning Environments Support in the UK Further and Higher Education Communities: Proposal for a New Support Service*, JISC Report 10/06/2002. [On-Line]. Available:
www.jisc.ac.uk/index.cfm?name=mle_related_les

Cunningham, S., Ryan, Y., Stedman, L., Tapsall, S., Bagdon, K., Hew, T., & Coaldrake, P. (2000). *The Business of Borderless Education*. Canberra: Department of Education, Training and Youth Affairs.

DFES (2003), *The Future of Higher Education*, London: Department for Education and Skills. [On-Line]. Available: www.dfes.gov.uk/highereducation/hestrategy

Greenalgh, T. (2003). E-learning expert castigates UkeU for "wasting taxpayers' money. *THES*, 9 May.

Jacobs, G. (1997). *The Dearing Report: A Summary with respect to Learning Technology* . Oxford: ALT.

Jacobs, G. (2001). The changing face of ALT-C, *ALT-J*, 9, 1, 2-16.

Keegan, M. (2000). *E-Learning: the Engine of the Knowledge Economy.* New York: Morgan Keegan.

Loncraine, R. (2002). Fancy some grown-up, intellectual chat on the net? *The Guardian*, 10 December.

Molyneux, S. (2003). Eyeing up the future. *THES*, 7 February.

National Committee of Inquiry into Higher Education (NCIHE) (1997) *Higher Education in the Learning Society.* [On-Line]. Available: http://www.leeds.ac.uk/educol/ncihe/

O'Leary, J. (2003). British higher education cannot take the chance that technology is simply another passing fad. *THES*, 7 February.

Phipps, R., & Merisotis, J. (2000). *Quality On The Line: Benchmarks for Success in Internet-Based Distance Education*, Washington DC: Institute for Higher Education Policy.

Robins, K.,& Webster, F. (2003). *The Virtual University? Knowledge, Markets and Management*, Oxford: OUP.

Smith, T. (2001). *Strategic Factors Affecting the Uptake, in Higher Education, of New and Emerging Technologies for Learning and Teaching*, JISC Technologies Centre. [On-Line]. Available: http://www.techlearn.ac.uk/NewDocs/HEDriversFinal.rtf

Squires, D. (1999). Peripatetic electronic teachers in higher education. *ALT-J*, 7,3, 52-63.

Tiffin, J., & Rajasingham, L. (1995). *In Search of the Virtual Class*, New York: Routledge.

Woodfield, R. (2003). Farewell to the duvet degree, *THES*, 23 January.

Notes

[1] http://www.fathom.com

[2] http://www.uoponline.com

[3] http://umuc.edu/gen/virtuniv.html

[4] http://www.u21global.com

[5] http://www.futuresproject.org

[6] http://www.allLearn.org

[7] http://www.londonexternal.ac.uk

10

Understanding Enthusiasm and Implementation: E-Learning Research Questions and Methodological Issues

Gràinne Conole

This chapter provides an overview of current learning technology research and demonstrates how it has emerged as a new field over the past decade. It will consider the cognate disciplines which feed into the area and the ways in which they inform it, as well as considering how these provide rich multiple perspectives but also give rise to inherent tensions due to differences in values and language. Current research activities are categorised, along with illustrative research questions. An outline of the research methods used to investigate these questions is given, along with an overview of the methodological issues that arise. The chapter reflects on how the area has developed over the past decade and concludes with some thoughts on future research directions. It describes the different dimensions of learning technology in terms of both the pedagogical and technical aspects and also the impact of technologies on both individual tutors and learners and on organisational structures.

The Emergence of Learning Technology as a Research Area

A review of research areas (e.g. Biochemistry or Computer Science) that have developed in the last hundred years or so, show a similar pattern of emergence, involving the stages outlined below. Rekkedal (1994) outlined a similar model for the emergence of distance education; in terms of mapping the terminology, definition, nature and focus of the field:

- *Pre-subject area*: no evidence of the area or perceived need or interest;
- *Beginnings:* individuals begin to research or ask new questions or issues arise which are triggered by some event or catalyst;
- *Emergence:* more researchers begin to work in the area and a community begins to develop;
- *Diversification:* the area starts to mature and different schools of thought emerge and the area begins to align or take place alongside more established areas;
- *Establishment:* the area becomes recognised in its own right with a defined community, experts, associated journals and conferences and is perceived of as 'respected' research with associated professional status, courses and career routes.

In terms of this model, learning technology research is currently between the stages of diversification and establishment. It is eclectic in nature, covering a broad church of research issues and is as yet not a rigorously defined area (Conole, Cook & Ingraham, 2003). A key tension is the struggle for recognition alongside established areas. This is linked to issues of shared dialogue and understanding for the area and articulation of the different schools of thought. However, learning technology has not arisen in isolation and feeds on a number of cognate disciplines; therefore research into technologies for learning per se has been an active area of interest with a long history. Despite this, Mason (2002) points out:

> Although e-learning builds on over 150 years of practice in distance education, it differs markedly from previous technological innovations and does not yet have an established research base. So far e-learning has not produced a new theory of learning; in its present form it can be analysed and interpreted using existing theoretical models. E-learning has, however, defined a new paradigm for learning; a way of working, studying and problem solving which reflects the growing connectivity of people and learning resources.

It is worth comparing the characteristics of learning technology with a well-established discipline like Chemistry, which has a long history (at least 250 years as a recognised field, when it emerged from Alchemy). It is divided into defined schools of thought (Inorganic, Organic, Physical) and has refined and sub-divided into new areas over time. It is characterised by a set of shared values and interests, with key research questions and foci of study. There are favoured methodologies and approaches. It builds on a substantive body of shared and validated knowledge. The community is fostered and developed through established conferences, journals and experts.

In terms of researching the use of technologies for learning, there was a paradigm shift (Kuhn, 1970) a decade or so ago. This was in part due to the substantive impact of the Internet on learning but was also fuelled by a number of national initiatives and policy drivers. A map of these initiatives and drivers and the kinds of research and development activities which emerged are discussed in more detail by Conole (2002) The research work and learning technology projects and developments which emerged as a result of these national initiatives in general lead to an increased interest in the role of technology across education, senior management engagement, and consequential change in strategy. This provided opportunities to experiment using developmental funding. Finally there was an influx of researchers and a growth of new 'centres' and expertise looking at learning technology. In parallel a series of dedicated conferences (such as ALT-C, EdMedia, Networked Learning, CAL) and journals (such as Computers and Education, JCAL, ALT-J) began to arise to foster the debate and development of the community, which shows that the area is becoming established.

This section has considered the factors that have influenced the emergences of learning technology as a research area and considered how established the area is alongside other research domains. It is clear that this is now a recognised albeit young research area. In the next decade or so we are likely to see the area continue to diversify but it is also likely that certain core research areas and foci of interest will emerge.

The Nature of Learning Technology Research

Learning technology as a research area is by its nature multi-disciplinary and covers a vast range of research topics, ranging from those that focus more on the technologies to those that focus on socio-cultural issues, the impact of technologies on learning and teaching, professional roles and identities, organisational structures and associated strategy and policy development. However, despite this diversity there are a number of common themes, which link the different research areas:

- Interdisciplinarity and multiple voices;
- Access and Inclusion;
- Change;
- Convergence and interoperability;
- Interactivity and social interaction;
- Politics.

Interdisciplinarity is a defining characteristic of the learning technology research area and is concerned with the influence of different research perspectives and also how problems in the practice of different disciplines result in differing adoption and use of learning technologies. Access and inclusion includes issues around the widening participation agenda, equity, access to technologies, barriers to inclusion and issues around the nature and extent of the digital divide. Change as a theme, focuses on exploring the drivers and rationale for learning technology focused change and their consequences and impact. It also addresses the strategies for managing and enabling change and mechanisms for implementation. Convergence and interoperability includes an exploration of different forms of convergence (technological, pedagogical, organisational, sectors and institutions) and also considers issues associated with scalability and globalisation and the underpinning standards needed to support interoperability. A current focus is also on a critique of convergence verses standardisation across the technical, pedagogical, human and organisational aspects. Interactivity and social interaction explores the interactivity of different tools and the nature of the medium. It also considers interactivity at different organisational levels and the ways in which organisational boundaries and functional groupings have blurred as a consequence of new technologies. Finally it considers the potential of technologies in terms of enhancing communication and collaboration and in building new communities and networks. Politics is a very strong theme that runs across all learning technology research. This in part relates to the over hyping which occurs, leading to an over expectation of what is possible. It is also partly due to different local agendas and associated infighting as well as the major impact that technologies can have. Whilst these different perspectives bring strength and depth to the learning technology research area, tensions also arise in terms of the different schools of thought and a lack of shared language.

Research Themes and Related Questions

The learning technology research area has developed significantly in the last decade and matured into a rich set of inter-connected research themes. This section outlines some of the current research themes and questions. Early work tended to focus on the technical aspects, with a focus on standalone multimedia applications and in particular navigational issues. In contrast there is now a broader base of research which has expanded partly because of the impact of the Internet and the ways in which it can be used to support learning and teaching, but also because of the increase of different learning management environments and systems. In particular there has been an expansion of research exploring the ways in which learning technologies can be used to support communication and collaboration, coupled with an increased focus on the associated pedagogical and organisational issues. This shift in research focus is evident in a review of articles published over the last decade. For example the first issue of ALT-J included papers exploring the relationship between cognitive styles and computer assisted learning, the design and development of multimedia and the use of iconic representation. In contrast, the latest edition has a much broader range of papers reflecting the expansion and maturing of the area. For example it reports on a range of national funding projects, the development of digital resources, analysis of approaches to study and experiences in online learning environment, profiling and understanding student behaviour and peer assessment of essays.

An account of the changing nature of the field of learning technology research, which provides a snapshot of how the area has developed, is given in 'The changing face of learning technology' which contains a selection of papers published in the journal ALT-J over the past decade (Squires, Conole & Jacobs, 2000). The selection was guided by four themes: design and evaluation of technology-mediated learning environments, institutional change, learning technology in a networked infrastructure and reflections on future possibilities.

Current research interests in learning technology can be grouped around three main themes: pedagogical, technical and organisational. These themes sit within a wider socio-cultural context that informs and influences the research agenda. The first of the three themes, focuses on the pedagogy of e-learning, and in particular the development of effective models for implementation, mechanisms for embedding the knowledge gained from learning theory into the design of learning technologies and their use in learning and teaching. This area also focuses on the guidelines and good practice to support the development of e-learning skills, the literacy needs of tutors and students, understanding the nature and development of online communities and different forms of communication (and associated issues of overload) and collaboration, different mechanisms for delivering and increasing flexibility and modularisation of learning opportunities and exploration of the impact of new emerging influences on learning, in particular the impact of gaming. This also covers the instructional aspects such as understanding effective design principles and promulgating good practice in the design and development of materials, exploration of different models for online courses, cultural differences in the use if online courses, requirements in terms of tutor support needs, time investments, mechanisms for improving student learning experiences and retention rates.

The second theme focuses on the underpinning technology of e-learning, including the development of the technical architecture to support different forms of learning and teaching, different mechanisms of monitoring and tracking activity online, exploration of the nature of different types of virtual presence, context sensitive, mobile and smart technologies and the hardware and software requirements. A lot of work has also been done on metadata, specifications and standards, interoperability, learning objects and most recently an educational modeling language, which aims to describe the learning process (see Chapters 2 & 3).

The third research theme explores issues which arise at an organisational level, including effective strategies for integrating online courses within existing systems, development of organisational knowledge, new methods and processes for developing a learning organisation and for the seamless linking of different information processes and systems. This has been a major focus for much of the recent work on Virtual and Managed Learning Environments (VLEs & MLEs). Lessons learnt and guidelines are currently being produced as an MLE development pack, which will be available from the JISC InfoNet service [1]. Table 1 outlines some of the typical research questions that are being addressed in each of the three themes.

The context in which research questions are being posed
The three themes outlined in the previous section sit within a context of factors and influences ranging from:
- Policy drivers & funding opportunities;
- National agendas and local initiatives;
- International research.

Table 1. From research themes to research questions.

Research focus	Types of research questions
Pedagogical aspects	What is effective pedagogy in using learning technologies? Are the new technologies potentially providing new forms of pedagogy? What are the best methods of integrating the use of new technologies within the broader learning and teaching context? What are students' experiences of using new technologies? What new forms of literacy are emerging for both students and teachers? What different cognitive support mechanisms do learning technologies offer? What are the inherent affordances of different technologies? How 'usable' are learning technologies and what factors influence their usability?
Underpinning technology	What are the new and emerging technologies and how can they be used to support learning and teaching? How can underpinning standards be developed and interoperability issues explored? Can we develop and test technical infrastructures and architectures? What are the emerging new software and hardware systems? What is the potential of mobile and smart technologies? How can we track and monitor materials use and navigational patterns? Can we develop an understanding of multiple forms of representation? How can images be used to support learning and teaching?
Organisational issues	In what ways can new technologies be used to support and enhance organisational learning? What mechanisms and procedures are there for developing shared knowledge banks of expertise and information? How do stakeholders (academics, support staff, students) currently work? How will current structures, skills and roles be mapped on new ones? What are the organisational issues and challenges associated with implementing an MLE and how will local contexts and issues influence this? How can the information flow processes for different types of activities, such as assessment processes, student registration, and course management be documented? What mechanisms for providing remote access to a variety of different users are there?

Funding opportunities have had significant influence on the direction and scope of research activities. Research in the UK for example has been heavy influenced by the Teaching and Learning Technology Programme funding streams and by the JISC programmes (see Chapters 4, 5 & 6), as well as funding programmes from more discipline-focused councils such as EPSRC and ESRC. Alongside the specific funding calls, policy drivers also have a significant impact. One example is the development of the National Grid for Learning across all schools in the UK, which has resulted in a host of related resource developments and training initiatives (see Chapter 5). Another is the directive by HEFCE for all Higher Education Institutions to produce detailed learning and teaching strategies.

Current local agendas and initiatives include access and accessibility issues, widening participation, gender issues, e-business and business modelling, technologies for lifelong learning, plagiarism, digital rights and IPR issues.

Research in the UK sits within the context of a wider body of international research. One area that is by definition international in scope is the standards and interoperability research. An outline of the international research context is outside the scope of this chapter but a reasonable overview can be found in the edited collection 'Distance Education – new perspectives' (Harry, Keegan & John, 1993) or two more recent collections (Steeples & Jones, 2002; Lockwood & Gooley, 2001). Research in different countries is shaped by the following factors; historical developments and the focus of previous research in the area, impact of national and international drivers and initiatives, funding, technological developments and specific local issues and agendas (see Chapters 7& 8).

Learning technology as a research area is rich, diverse and complex and to understand this there is a need for a variety of perspectives and methodological approaches to be undertaken. The area draws on the literature from across a broad spectrum of research fields from educational research, general social science and business studies through to computational science, mathematical modelling and linguistics. Furthermore the nature of the medium being studied also raises complex and new methodological issues. However, as part of the process of becoming established the area will need to demonstrate that the research being carried out in the field is methodologically rigorous, building appropriately on existing knowledge and theories from feeder cognate disciplines.

Research Methodology Issues

Learning technology research in general is concerned with understanding how technology can or could be used to support learning and teaching with an underlying motivation of improving the student learning experience, as well as exploring the impact of learning technologies from an individual through to an institutional level. To achieve this it is worth reflecting on which research methodologies might be appropriate to address the questions outlined in Table 1. The choice of methodologies and the way in which it is carried out in terms of the data collection and analysis will have a critical impact on how well these questions are answered. This section provides a review of some of the current methodological arguments and issues, such as consideration of the validity of the research design, and the rigorousness of the research methods, data collection and analysis.

Essentially there are two broad classifications of research carried out in this area: evaluation and research studies. Evaluation studies involve both formative and summative evaluations, and include case studies and individual investigations. In contrast, research studies can be considered to build on underpinning theoretical approaches to provide a rationale for findings and results.

Evaluation studies
One of the defining characteristics of evaluation is that it is often research work
commissioned by particular stakeholders to address specific questions. As such, one
of the important facets of evaluation is stakeholder analysis and a clear identification
of their interests in terms of the evaluation. Focusing on different stakeholders will
have a direct impact on the scope and nature of the evaluation from the design
methodology through to the data collection and analysis and even presentation of
findings. A number of evaluation frameworks for learning technology have been
developed, such as the SECAL framework (Gunn, 2001), the integrative framework
(Draper et al., 1996), and the Evaluation of Learning Technologies (ELT) framework (Oliver, 1997). Evaluation has become increasingly important in terms of understanding the development and use of learning technologies across the broad spectrum of research questions outlined above. In essence, evaluation studies can have a
range of purposes; from selecting, or monitoring through to researching or providing
evidence. Oliver defines evaluation as the process by which people make value
judgements about things. He goes on to state that in the context of learning technology, these judgements usually concern the educational value of innovations or the
pragmatics of introducing novel teaching techniques and resources (Oliver, 2000).
Given the complexity of the research domain and the highly political nature of the
area, evaluations need to be pragmatic in nature, focussing clearly on appropriate
stakeholder needs. This approach is central to Patton's utilization-focussed approach
to evaluation, which considers evaluation as a mean to an end, rather than an end in
itself (Patton, 1997). Support for practitioners, in terms of developing an evaluation
plan and conducting evaluation research, is provided by the Evaluation Cookbook
(Harvey, 1998) and the Evaluation Toolkit [2]. The latter helps practitioners, irrespective of their current degree of expertise, to evaluate the use and range of learning
technologies. It provides a structured resource that can be used to plan, scope and
cost an evaluation (Oliver et al, 2002; Conole et al, 2001).

Research Studies
The choice of appropriate research methods will depend both on the nature of the
questions being considered and on the associated stakeholders in the research findings, as stakeholders may have conflicting agendas and are likely to place different
values on methodological approaches. Broadly speaking there is a tension between
the needs of policy makers and senior managers and academics and support staff on
the ground level. The former group are much more likely to be interested in potential efficiency gains and cost effectiveness associated with learning technologies and
will want to see evidence-based practice with a comparison of the benefits of new
technologies over existing teaching and learning methods. Oliver and Conole argue
strongly against this push towards evidence-based practice in e-learning, (Oliver and
Conole, 2003) stating that:

> Policy makers are increasingly looking to evidence-based practice as a
> means of ensuring accountability and validity in education and more
> recently in e-learning [...]

But that this results in:

> a number of implications for e-learning, including the problems facing practitio-
> ner-researchers working on project funding and the potentially distorting effect
> of e-learning policy on research in this field.

Oliver and Conole (2003) review the origins of evidence-based practice and consider this in relation to the emergence of e-learning as an area of policy, research and
practice and argue that this raises a number of methodological, epistemological and
moral questions. In contrast to policy makers, practitioners are much more likely to
be interested in exploring the properties and affordances of learning technologies,
ideas for how these can be used to improve the learning experience, gaining an understanding of the associated issues and impact of new technologies on different
stakeholders, organisational structures and the wider learning and teaching context.
One of the issues raised by Oliver and Conole is the emphasis evidence-based practice places on quantitative research methods over qualitative ones.

As a relatively young field, learning technology research suffers in a number of
respects. Firstly, the area is not yet clearly defined and scoped. Secondly, there is
considerable criticism of much of the current research activities, as it is considered
too anecdotal, case study based, and lacking theoretical underpinning. Thirdly, as
indeed is true in social science research more generally, there are divided views on
the importance of quantitative versus qualitative research methods. However, Hammersley points out (1997) that a preference for quantitative methods in educational
research is inappropriate stating that both quantitative and qualitative methods have
roles to play in the process of research. Oliver and Conole (2003) agree with this and
in the context of learning technology state that:

> This is particularly true in the field of e-learning which, being a relatively new
> field of study, remains contested by the various disciplinary traditions (educa-
> tion, psychology, computer science, etc.) that contribute to it. This contestation
> means that no single model has arisen to explain how e-learning works; psycho-
> logical theories such as constructivism sit alongside social scientific theories
> such as the notion of habitus and cognitivist theories such as cognitive load the-
> ory (Oliver & Aczel, 2002). Lacking any singular model, it is impossible to rec-
> oncile the diverse studies that are undertaken in any systematic way. Each has to
> be interpreted on its own merits and reconciled with other studies in a way that is
> sensitive to the theories involved. As such, it has been argued that all studies in
> this field should, methodologically, be interpreted as case studies, even when
> they adopt experimental approaches (Holt, McAvinia & Oliver, 2002).

Mitchell (2000) takes a different perspective, condemning the lack of research rigour
in learning technology research in his article 'The impact of educational technology:
a radical reappraisal of research methods'. He argues that few of the papers in this
area meet all or even most of the requirements of scientific research (theory-
orientated, conceptually clear, measurements consistent with number and measure-

ment theory, preconditions for parametric statistics met, appropriate logic, replicative validity). He makes some suggestions for addressing this. Firstly research students should receive training in which they are made aware of literature from feeder cognate disciplines, as well as epistemology and the philosophy of science. Secondly he argues that the first stage in educational research should focus on epistemological rather than statistical or design issues. Thirdly he points to the role of journal editors in this field in terms of tightening up refereeing procedures and engaging in more critical feedback with authors on the quality of their submissions. He suggests the 'Statistical Guidelines for Contributors to Medical Journals' (Gardner & Altman, 1989) as a useful reference in this context.

Methods and Approaches

This section will give examples of common data collection methods, which are used in learning technology research and ways in which particular approaches might be of value in specific circumstances. Data collection is influenced by the very character and nature of the research area (multi-faceted, complex, and dynamic) and associated influences (political nature, different stakeholder agendas). Furthermore in general the research is usually pragmatic in nature, seeking to understand the answers to real research questions linked in particular to exploring how learning technologies can be used to improving learning and teaching. The section is not intended as a detailed critique of research methods as there are numerous texts of this type (e.g. Cohen & Manion, 1996; Frankfort-Nachmias & Nachmias, 2000).

Surveys and questionnaires are a standard means of data collection used across the different threads of learning technology research. They have the advantage of being easy to collate, with simple response and closed question types being amenable to quantitative analysis. However good questionnaire design is a specialised skill and requires considerable investment if there is any hope of meaningful results. Furthermore, there are often low response rates to questionnaire calls. One final point is the importance of careful targeting of respondents, which can be particularly difficult in our area of research where learning technology experts are located in different parts of institutions across the sector with a wide range of roles and job titles from those which focus centrally on learning technology through to more traditional academic roles with a learning technology component (Beetham, Jones & Gornall, 2001).

Interviews with key stakeholders can be a valuable mechanism for both initial fact-finding and ongoing analysis of stakeholder views and perspectives on the projects. It is important to spend some time establishing who key stakeholders might be and their interest in the project outcomes. However, interviews can be both time consuming in terms of arranging interviews schedules, conducting the interviews and then analysing the results. It is important to be clear about both the purpose and the format of the interviews.

An alternative to interviews is to carry out a series of focus groups of representative stakeholders. As with interviews, it can be difficult to find an appropriate time to get users together for the focus group. A range of strategies may need to be employed for organising focus groups in a way that is confluent with other demands on users' time. Both interviews and focus groups can be used to provide a range of views about the organisation, political and economic issues facing the university and the challenges and opportunities it faces. With this respect, it is useful to build up different models of the nature of a university to capture these different aspects.

Another technique for gathering data is observation; this can be used in a variety of circumstances such as observing the use of new software to establish usability issues or exploration of general navigational issues, participation in workshops or the use of materials to establish their effectiveness or any emergent issues. Data can be gathered by the researcher taking notes on specific activities or outcomes or can be supported by audio or video recording the session. However, it is worth noting that the latter adds considerably to the analysis time in terms of transcribing and interpreting the recordings.

Process mapping can be a useful means of identifying existing structures and practices. One mechanism for achieving this is to build a systems picture of the university, having already obtained information about the management, service and department structures and processes and then use this as a means of developing a model. This technique has been used extensively, for example, as a precursor for establishing user requirements and specifications for MLEs, but also in the development of specific software systems, web sites and portals for specific purposes.

Project working groups, meetings and committees can also be a valuable source of information. Analysis can include interpretation of the discussions held during the meeting, as well as the documented meeting minutes. Using pseudo-focus groups for feedback on particular aspects of the project might be considered. This is a useful way of addressing the issue of the difficulties of getting different stakeholders together.

Analysis of documentation can provide valuable insights into existing practices and structures. A variety of sources can be used such as minutes of meetings and committees as discussed above, but also project plans and reports, departmental documents, strategic and operational plans, policy documents, handbooks and guidelines. In addition organisational charts and functional charts and role specifications are valuable data sources. These can provide a snapshot of current activities and also a comparison of how structures and functions have changed over time, which can provide insights into the impact learning technologies have had and the ways they are being embedded into learning and teaching and also the impact on organisational structures, roles and functions.

Beetham and Conole have previously described the importance of the development of a sense of shared ownership and co-participation in learning technology research, in part because of the practical pragmatic dimension to the area but also because of the highly political nature and dependences on stakeholder perspectives (Beetham, 2001; Beetham, & Conole 2001) One of the studies they describe is the development and use of an instrument to analyse the nature and distribution of learning technologists in institutions. The method they employed was to recruit a representative sample of 'institutional auditors' who were then actively involved in the process of instrumental development. This was developed through a series of workshops, which included guidance on the ways in which the study could be used by the auditors to address local issues within their own institutions. Guidance was also provided on the different means of collecting information across the institution. As a result of this co-participative approach auditors were much more engaged with the process of data collection and had a shared sense of ownership and responsibility. This meant that a much richer set of data were collected, which were therefore also more authentic. The cumulative results provided a benchmark of indictors for different institutional factors allowing auditors to position their institution in terms of learning technology against a national 'average'.

Data Analysis and Dissemination

This section is not intended to provide a detailed outline of different analytical techniques but rather it will pick out examples and approaches that are commonly used in learning technology research. In terms of interpretation of transcripts gathered from focus groups and interviews then either pre-coding classifications or a grounded theory (Strauss, 1987) approach is generally adopted. A range of quantitative methods can be applied to material gathered through closed questionnaires or via systems usage and Web log statistics, using a host of standard statistic tests and output mechanisms. A useful introductory text for this area is 'Learning to use statistical tests in psychology' (Greene & D'Oliveira, 1982). Similarly there are a host of software tools that can be used to collate, manage and analyse such results, the most widely used being SPSS. Similarly for analysis of qualitative data there are a range of software tools for managing and analysis such as NVIVO, which helps in terms of coding, classification and analysis.

Interesting methodological issues have emerged in terms of gathering and interpreting data using learning technologies, or in researching their use. A good example comes from the work that has focussed on researching the nature of online discussion forums and in particular the ways in which they support student learning, communication and collaboration. Early work in this area focussed primarily on simple analysis of the content of the threaded messages. There was a naïve assumption that this was enough to capture the whole event, without an understanding of the context within which the discussion took place Analysis typically used pre-defined coding schemes, such as those developed by Henri (1994). However these pre-defined classifications have been criticised for being too prescriptive. As an alternative, some researchers opt to develop their owning coding schemes tailored around their specific research interest. Others choose to explore emergent themes. The pros and cons of each of these approaches has been hotly contested and debated, as have the different mechanisms, which are then used to analyse the results. Furthermore, the interpretation which can be placed on these findings has been contested mainly because there is a consensus that a focus solely on the outputs of the discussion forums only provides a partial view of what is actually happening, in particular because it ignores the socio-cultural context within which the event took place (McConnell & Hodgson, 1999; Lally & De Laat, 2000). Discussions as to what analysis approach can be used to address this have tended to favour a multi-modal data collection process, which provides a richer description of the event. This mirrors much of the current movement in educational theory and in particular the school of thought around

around the use of Activity Theory as a unit of analysis (Engestrom, Miettinen & Punamaki 1999).

There are a variety of ways in which research data can be used. For example, it can provide a mechanism for undertaking an audit of current activity, such as a snap shot of learning technology activities across an institution, to provide an indication of the degree of uptake of a VLE across different subject domains. It can also be used to map learning technologies against structures and functions or as a means of monitoring change and benchmarking over time.

Having carried out an evaluation or research study and analysed the results it is important to give careful consideration of how and where to report on the findings and to map potential audiences and stakeholders. There are a variety of dissemination mechanisms and outlets from peer-reviewed academic journals through to presentations and oral reports. In terms of journals there are essentially two main types, those that have a particular focus on learning technologies (such as ALT-J, JCAL and Computers and Education) and subject specific journals (such as Chemistry Education). Table 2 has been taken from the presentation section of the Evaluation Toolkit discussed earlier [2]. It outlines potential dissemination outputs and suggestions of when they might be used.

In terms of the structuring of different types of articles for ALT-J based on the nature of the research, Oliver has produced some helpful guidelines and templates, which are available online on the Association for Learning Technology web site [3]. There are templates for reporting on four different types of study; a study that has carried out a survey, a report of a literature review, a particular critque of an aspect of the area or specific theoretical position, and a study reporting on development and testing findings.

The Importance of Learning Technology Research

There are a number of reasons why learning technology has become an important research area. Firstly learning technologies now have a significant impact at all levels of universities and colleges, including organisational, structural and functional as well as teaching and learning. However, little is still understood about these processes and how they are changing. There is a need to research these in order to draw out lessons learnt and examples of good practice, which can then inform future developments. Secondly, the variety and complexity of new technologies and the potential ways in which they can be used is changing rapidly. Thirdly, partly because of the first two factors, more and more academics and support staff are now becoming involved in learning technology research as part of their roles or as a means of understanding how learning technologies can be used effectively.

Table 2: Mapping dissemination vehicles to functions.

Dissemination vehicle	Function
Peer-reviewed journal	Detailed research-orientated structured report/paper.
Newsletter	Regular update of project outputs and short news stories
Conference Presentation	Presentation of findings at a conference, usually adapted/tailored towards aspects of the conference theme
Web site	Structured set of resources or information.
Promotional leaflet	Short summary guide or dissemination paper-based format.
Email	Online asynchronous one-to-one or targeted one-to-many communication/dissemination.
Press release	Semi-formal summary of main findings usually for a targeted audience.
Poster	Used primarily at conferences. Usually consists of a series of A4 or A3 laminated sheets
Report	Project progress or summary report about a particular aspect of the work
Word of mouth	Verbal presentation or update or results, may be used in a range of forum, from informal one-to-one, through departmental meetings or at more formal steering group or external meetings.
Workshop	Either for dissemination or training/teaching purposes.
Presentation	Verbal presentation usually supported by a report/PowerPoint presentation or summary notes/overheads. Can be used both for in-house meetings (departmental, project, etc) and external meetings (conferences, workshops, etc.)

If learning technology is now an important research area then it needs to be recognised as valid and given kudos alongside more established research areas. This is particularly important in terms of encouraging academics to become involved in learning technology research alongside their mainstream subject-based research. If the learning technology research is not valued and recognised as equal then it is unlikely that academics will be prepared to invest time in this or be prepared to divert effort away from credible 'real' research work. This is particularly problematic in institutions that have a strong drive toward being considered as research-led institutions.

Credibility in learning technology as a research area is also important in terms of educating senior management about the complexity of the area, to help inform the decisions they need to make in terms of different ways in which learning technology impact on their business. Otherwise there is a danger that they will make ill-informed and rash decisions based on scant evidence and/or will be beguiled by hype rather than understanding the true complexity of the domain space. To date, there have been a number of examples of where this has occurred. One example is the purchasing of a VLE to support learning activities and then decreeing that all courses must use the system without considering whether or not this might be pedagogically appropriate or thinking through the associated staff development needs and time implications. Similarly, senior managers frequently demand evidence of the cost effectiveness of the use of a particular technology without thinking through whether they have a genuine understanding of the comparative costs of traditional teaching methods or indeed whether a direct comparison is actually possible given that the introduction of the technology may result in a significant change in the learning and teaching process (see Chapter 9 for a consideration of the decisions senior managers make regarding developing global e-learning markets)

Likely Trends in Learning Technology Research

It is dangerous to try and predict what future directions and research activities are likely to predominate in a technologically fast moving area; however a few general observations about likely trends can be made. Firstly, it is likely that as the area matures better use and application of relevant theories and models from cognate disciplines will be applied and more examples of context-specific theories and models will emerge. In other words the area will shift from a focus on anecdotal case studies to an area more grounded in a theoretical approach. Secondly, as learning technologies become more embedded in institutions and as larger cross-institutional initiatives are put in place there will be more evidence of the factors and issues involved in mainstreaming and scaling up. Thirdly, the pace of change in terms of new technologies is not likely to slow; in particular mobile and smart technologies are likely to have dramatic effects. The study of these technologies offers new challenges and will need to draw strongly on the knowledge developed in Psychology, Artificial Intelligence and related fields. Whatever the research activities are, what is certain is that this is a fast moving and rapidly developing area, which will continue to mature and diversify in the coming years.

Conclusion

This chapter has provided an overview of learning technology as a research field. It has described the emergence of the area over the past decade, highlighting the characteristics, which define the area, the impact of feeder cognate disciplines and the inherent tensions and multi-perspectives. A summary of current research activities is described along with key questions of interests. Common research methods deployed in the area are described along with some of the epistemological and methodological issues, which arise.

I have tried to demonstrate that learning technologies are now having a significant impact at all levels of an organisation. This is reflected both in the current research focus and questions, the majority of which are exploring broad organisational/structural issues rather than specific local or individual concerns. There has been a shift away from isolated small-scale studies such as whether a particular technological intervention will work with students on a specific course with a particular department. This shift is reflected in the research methodologies that are used to explore the wider political and organisational context in which learning technologies are used. The development of learning technology research as an accepted field and discipline also reflects a move away from the 'individual enthusiast' working in their own subject discipline to a community of connected enthusiasts developing an exciting and dynamic discipline. Furthermore there is now more recognition of the importance of the area, providing insight into the way technologies are fundamentally altering both individual practice and organisational structures and processes.

It is hoped that this chapter gives a flavour of the richness of learning technology as a research area and a sense of the excitement of working in this fast moving field but also a feeling for the associated challenges and methodological issues. The next decade will be critical in terms of the area finding a clear niche and position alongside more established research fields. Then there is a real potential that the research findings will begin to offer us some real insight into the ways in which technologies can effectively support learning and teaching and an understanding of how they can be used to improve organisational processes. Hopefully we will also begin to see the development of new underpinning theories and models of explanation to account for the use of learning technologies and perhaps even the emergence of new learning paradigms and working practices. Only time will tell.

References

Beetham, H. (2001). *Developing learning technology networks through shared representations of practice.* In Proceedings of the 2001 9th International Symposium Improving Student Learning: Improving Student Learning Using Learning Technologies, Oxford: OCSLD, Oxford.

Beetham, H., & G. Conole (2001). *Modelling aspects of institutional developments: culture, infrastructure, expertise.* Improving Student Learning Using Learning Technologies Conference, Heriott Watt, University, Edinburgh.

Beetham, H., Jones, S., & Gornall, L. (2001). *Career Development of Learning Technology Staff: Scoping Study, a final report for the JISC JCALT.* Bristol: University of Bristol.

Cohen, L.,& Manion, L.(1996). Research Methods in Education (4th Edition). London: Routledge.

Conole, G. (2002). The evolving landscape of learning technology research. *ALT-J* 10,3, 4-18.

Conole, G., Cook, J., & Ingraham, B. (2003). *Learning technology as a community of practice.* Research Paper Presented at ALT-C 2003.

Conole, G., Crewe, E.T. Oliver., & Harvey, J. (2001). A Toolkit for Supporting Evaluation. *ALT-J*, 9,1,38-49.

Draper, S., Brown, M.I., Henderson, F.P., & McAteer, E (1996). Integrative Evaluation: an emerging role for classroom studies. *Computers and Education*, 26, 1-3, 17-32.

Engestrom, Y., Miettinen, R., & Punamaki, R.-L.(1999). *Perspectives on activity theory.* Cambridge: Cambridge University Press.

Frankfort-Nachmias, C., & Nachmias, D. (2000). *Research Methods in the Social Sciences*, 6th ed New York: Worth Publishers.

Gardner, M.J., & Altman, D.G. (1989). *Statistics with confidence*, London: British Medical Journal.

Greene, J., & D'Oliveira, M. (1982). *Learning to Use Statistical Tests in Psychology.* Milton Keynes: Open University Press.

Gunn, C (2001). Effective online teaching- How far do the frameworks go? In G. Kennedy, M. Keppell, C. McNaught & T. Petrovic (Eds.), *Meeting at the Crossroads.* Proceedings of the 18th Annual Conference of the Australian Society for Computers in Learning in Tertiary Education. (pp. 235-244). Melbourne: Biomedical Multimedia Unit, The University of Melbourne. [On-Line]. Available: http://www.ascilite.org.au/conferences/melbourne01/pdf/papers/gunnc.pdf

Hammersley, M. (1997). The relationship between qualitative and quantitative research: paradigm loyalty versus methodological eclecticism. In J. Richardson (Ed.) *Handbook of qualitative research methods*, (pp 159-174). Leicester: The British Psychological Society.

Harvey, J. (1998). *The LTDI Evaluation Cookbook.* Edinburgh: Learning Technology Dissemination Initiative.

Harry, K., Keegan, D., & John, M. (1993). *Distance education: new perspectives.* London: Routledge Falmer.

Henri, F. (1994), Distance learning and Computer-Mediated Communication: Interactive, Quasi-interactive or monologue? In C. O'Malley (Ed.), *Computer Supported Collaborative Learning* (pp. 145-161). Berlin: Springer-Verlag.

Holt, R., McAvinia, C., & Oliver, M. (2002). Evaluating Web-based learning modules during an MSc programme in dental public health: a case study. *British Dental Journal*, 193,5, 283-286.

Kuhn, T. (1970). *The structure of scientific revolutions.* London: The University of Chicago Press, Ltd.

Lally, V., & de Laat, M. (2000). *Cracking the code: learning to collaborate and collaborating to learn in networked learning.* In Proceedings of the Networked Learning Conference, April 2000, Sheffield.

Lockwood, F., & Gooley, A. (2001) *Innovation in open and distance learning - successful development of online and Web-based learning,* Kogan Page, London.

Mason, R. (2002). E-learning and the eUniversity. ALT Policy Board, Birmingham. [On-Line] Available: http://www.alt.ac.uk/docs/RobinMason.ppt

McConnell, D. and Hodgson, V. (1999), 'Methods for researching networked learning', *Proceedings of the 9th Improving Student Learning conference*, September 2000, Herriott-Watt University,

Mitchell, P.D. 2000. The impact of educational technology: a radical reappraisal of research methods, *Alt-J*, 5/1,p.48-54.

Oliver, M. (1997). A framework for evaluating the use of learning technology. BP ELT report no 1, University of North London.

Oliver, M. (2000) 'An Introduction to the evaluation of learning technology' in *Educational Technology & Society*, vol. 3, 4, 20-30.
http://ifets.gmd.de/periodical/vol_4_2000/intro.html

Oliver, M., McBean, J., Conole, G.,& Harvey, J., (2002). Using a Toolkit to Support the Evaluation of Learning. *Journal of Computer Assisted Learning*, 18, 199-208.

Oliver, M., & Conole, G (2003). Evidence-based practice and E-Learning in Higher Education. Paper presented at the Annual Meeting of the American Educational Research Association. Accountability for Educational Quality: Shared Responsibility April 21-25, 2003 [On-Line]. Available:
http://www.tigersystem.net/aera2003/revviewproposaltext.asp?propid=3397

Oliver, M., & Aczel, J. (2002). A commentary on the use of theory in the analysis of the Jape study. *Journal of Interactive Media in Education*. [On-Line]. Available: http://www-jime.open.ac.uk/2002/3

Patton, M.Q (1997). *Utilisation-focused evaluation: the new century text.* 3rd Ed. California: Sage Publications.

Rekkedal, T. (1994). *Research in distance education - past, present and future.* [On-Line]. Available: http://www.nettskolen.com/alle/forskning/29/intforsk.htm

Squires, D., Conole, G.C., & Jacobs, G. (2000). *The Changing Face of Learning Technology.* Cardiff: University of Wales Press

Steeples, C., & Jones, C. (2002). *Networked learning: perspectives and issues,* London: Springer.

Strauss, A.M. (1997). *Qualitative Analysis for Social Scientists.* Cambridge University Press

Notes

[1] http://www.jiscinfonet.ac.uk/).
[2] Evaluation Toolkit: http://www.ltss.bris.ac.uk/jcalt/
[3] http://www.alt.ac.uk

11

Looking Backwards, Looking Forwards: An Overview, Some Conclusions and an Agenda

Martin Oliver

The previous chapters in this book have explored the processes through which learning technology has been embedded in post compulsory education over the last decade. This chapter is somewhat different. In part, it is a review of those reviews, identifying themes and commenting on the analysis. It is also intended, however, to be forward-looking, learning from these reflections, thinking about how this might affect work yet to happen.

However, futurology is a dangerous game. Rather than starkly asserting that this or that will be the future, I intend to play safe by identifying likely trends. What I will add, however, are thoughts about what the future *should* be like. Inevitably, this will be a personal agenda, but rather than seeing it as the last word, uncontested, acting as the closing point of this book, I intend it to be an opening, the starting point for a new debate and an invitation to practitioners and researchers to take their own stand on how things ought to develop.

A Framework for Discussion

I have used three questions to frame my analysis of the previous chapters in this book:
* What have we learnt from the past?

- Are we questioning the present?
- Have we considered our role in this?

What have we learnt from the past?

A cornerstone of scholarship involves using previous activity to guide current practice (Boyer, 1990). Learning technology often seems an amnesiac field, reluctant to
cite anything 'out of date'; it is only recently that there has been a move to review
previous practice, setting current developments within an historical context (for example see Conole, 2003; Thorpe, 2002). Partly, this tendency to forget can be explained in terms of the speed of technological development. Nonetheless, many lessons learnt when studying related innovations seem lost to current researchers and
practitioners.

Are we questioning the present?

Much work in the learning technology field takes places under the auspices of short-
term funding, typically on projects with a developmental remit and pre-specified
outcomes (Beetham, Jones & Gornall, 2001). This does not place learning technologists in a strong position to criticise or challenge the remit they are given; typically,
they are beholden to funders or required to work within the limits of a pre-defined
strategy. This is a problem for a field that is now recognised as being inherently political (e.g. Jones, 2002). Barnett (1994) contrasts two notions of critique: one focusing on operations and seeking to increase efficiency, the other focused on understanding. When learning technologists are contractually bound to implementing pre-
specified projects, the first of these is emphasised: efficiency, usually in development or 'delivery' of courses. This leaves little scope for critiques of motives or assumptions. If learning technologists are unable to question the political motivations
behind the work they are asked to do, they become a tool.

Have we considered our own role in this?

The first two questions I have posed highlight whether learning technologists reflect
upon their work. Whilst this is a good beginning, arguably, it is not the whole story.
Further insight is to be gained by asking how we (individually and collectively) are
implicated in the stories played out around us. We need to consider our role reflex-
ively, not simply reflectively – we must analyse our own motives and practices, as
well as those we work with and work for.

   Lisewski & Joyce (2003), for example, question whether learning technologists
are attempting to legitimate their role by aligning themselves with dominant managerialist agendas, in effect, allowing themselves to become an unquestioning tool
so as to secure their own position, or else by 'going native' and becoming just another academic specialism. What position is taken within these chapters on the process of legitimating learning technologists' pedagogic expertise? Bound up with this
question, who has benefited from these initiatives, and who has been marginalised?
Considering questions such as these may provide us with the insights we need to
respond appropriately to developments and to groups in the future.

Emerging Themes

Six key themes emerge from my review of the chapters in this book:
- Cultures and change;
- Policy and Practice;
- The invisibility of embedding;
- Failure, maturity and theorising practice;
- Teaching, learning, content and technology;
- Treating teachers as learners.

Cultures and change

Our understanding of institutional change appears partial and confused. Perhaps because of the managerialist credentials learning technologists use to justify their power (Lisewski & Joyce, 2003), the analyses of the last decade tend towards a rational, technical and financial model of change. It seems strange, for example, that Jacobs presents student culture (and the valuing of social life) as a fundamental stumbling block whilst elsewhere asserting that "the main institutional force for a change of direction will not be pedagogical, but hard-boiled educational politics entrenched in matters of finance" (Chapter 9). Wilson suggests that the situation might be simpler in further education, where the ethos has been "conditioned" by funders to be competitive and expansionist – but even there, the dominance of economics may be a matter of degree rather than the sole cause of change (Chapter 5).

There are many alternatives to rational models of change, and it is interesting to see these emerging within other chapters. Dempster & Deepwell, for example, address the messy, contextual nature of change, the central role of debate throughout this process, the history shaping engagement and the tendency of projects to view phases that might contribute to success (such as implementation, evaluation, dissemination) "in a holistic and non-linear manner", rather than as a series of problems to be solved (Chapter 4). Such qualities are prominent in contemporary theories of change (e.g. Marsick & Watkins, 1999), validating their description of 'successful' projects which caused change without being able to 'tick the boxes' of their initial, rational plans. Rather than being bound to pre-judged outcomes, project staff responded, adapted – learnt from the complex, unpredictable situations they found themselves in , and had all the more impact as a result, even though, in Dempster & Deepwell's terms, "the influence of the project is sometimes hard to disentangle from general trends and movements in the institution", or only appears to have a "coincidental" impact on institutional policy and practice.

This informal, serendipitous kind of learning is apparent within Wilson's chapter too, particularly in the way that projects, relationships and developments "have been pulled together either at or on trains to and from ALT events". Learning Technologists seem particularly good at this kind of networking (Oliver, 2002), which might explain the dynamic evolution of the field. This also explains why, as Dempster & Deepwell argue, "the criteria against which projects are deemed successful within the project community itself relates to the process rather than the outcome", particularly in terms of collaborations fostered with other institutions or communities.

Whilst rationalism has its limits, we should not blind ourselves to the ways in which funding projects does lead to change. It buys time in which discussion, debate and learning can take place. Less clear, but equally important, it that sends a message: that these activities are valued, that if you wish to be seen as 'good' you should work this way. What is telling, however, are the relative priorities of funding councils in Scotland and the rest of the UK. Wilson's description of the activities of SCET, the Scottish Learning Network and the Virtual Learning Centre, for example, are striking for their inclusion of community-developed tools and professional development; Nora Mogey's case study in Littlejohn & Peacock's chapter describes the equally thoughtful approach of the Learning Technology Dissemination Initiative (see Chapter 6). Comparable qualities were praised in the EFFECTS project in later years, but it took two phases of TLTP before HEFCE realised developing resources was insufficient to shift teaching practice.

Echoing the research of Taylor (1999), Jacobs illustrates that change is not simply a neutral issue: it has a profound impact on academics' sense of identity. If technology is seen as a threat to the moral mission of academia, is it surprising that academics resist it? But why is it believed that technology threatens this moral mission? Jacobs links e-learning to for-profit initiatives. If this association is also made by academics, the moral threat becomes clear: e-learning becomes a way of achieving an economic good, not a social good. Appendix 2 of the Dearing report (NCIHE, 1997) provides one example of policy makers advocating such links and the recent White Paper builds upon this (DFES 2003). E-learning is *not* necessarily linked to financial imperatives for post-compulsory education; however, it has become *implicated* in these through government policy and managerialist discourses of efficiency and control (Holley & Oliver, 2000). Unless learning technologists actively challenge such rhetoric, academics are unlikely to spontaneously view the two agendas as separate and stop treating e-learning as a threat. Such academic resistance is brought to the fore in Littlejohn & Peacock's chapter. Programmes of development often focus on re-skilling and fail to address anxieties about role change and the loss of familiar practices. Faced with such potential loss, it is unsurprising that established academics are reluctant to engage.

It is interesting that students around the world are resisting e-learning, too. Oliver, O'Donoghue and Harper describe how students rejected the ideology behind resource-oriented learning, in spite of continuously improving material (Chapter 8). Similarly, Jacobs describes the growing phenomenon of students rejecting e-learning in favour of 'real' teaching and learning. Partly this might be prejudice, based on expectations rather than informed decisions. However, it is also a sensible choice: typically, students will have had years of face-to-face schooling and little or no experience of online learning. Why suddenly opt for a medium in which they are, in effect, culturally illiterate over one where they have learnt the rules and know how to judge their progress through the course?

This begs another question: in whose interest is the introduction of e-learning? Here, we see economic arguments for it and student arguments against. The picture is, of course, much more complex, but those tasked with implementing e-learning should ask whose ends are being served, and those who research it should consider who is given a voice to speak for or against these innovations.

Policy and practice

Recurrent throughout this book is the influence of policy on the work of learning technologists. Significant landmarks, such as the Dearing report, are visible as factors that have shaped practice. Funding followed this to set in train the UK's e-Universities; it also prompted a swathe of project funding. Indeed, this one report was singled out as being of central significance in shaping the field within the UK (Conole, 2003).

This situation is worrying: the Dearing report's discussion of the relationship between technology and Higher Education was primarily economic. For all its rhetoric about learning, students are described passively, at best placed as "informed consumers" and at worse completely absent from discussions about technology and learning (Smith & Oliver, 2002). Where is there explicit evidence that such perspectives are resisted? Jacobs even opens by asserting that Dearing gave learning technologists the authority to change the direction of work in this area and asserts that this vision is "destined" to be realised, in spite of recognising the lack of evidence for cultural change and the failed projects based on this collaborative computer-mediated distance course approach. Dearing certainly advocated a position that empowered learning technologists, and it is understandable that we are, collectively, attached to it, but we should not let our collective gain blind us to its shortcomings. This uncritical position on policy is indicated in Calverley's chapter (Chapter 2):

> Resource allocation can be aligned (people, tools, cost) to the new provision (risk) and application of the new policy. The process can then be controlled along the already-identified tension lines in a managed way, rather than spiralling out of control.

Given the complexity of change discussed above, where is the room for learning, for surprise, here? This is, doubtless, an *efficient* way to work – but it is about control, about people being made to work in pre-specified ways. What of their expertise, their perspective? Such a model can deliver short-term gains against pre-specified measures, but will it lead to the shared learning and organisational evolution so important to survival in the educational climate (Marsick & Watkins, 1999)?

Reassuringly, there is also evidence of policy being critically scrutinised. Jacob's analysis of the recent White Paper, for example, is refreshingly open in its assessment of the political sub-text that seeks to divorce teaching from research. Comparably, Oliver, O'Donoghue and Harper discuss the issue of whose responsibility funding for Higher Education is, highlighting the financial motives behind policies promoting e-learning. Moreover, there are clearly cases when policy should be rallied to, rather than challenged – such as Oliver, O'Donoghue and Harpers' example of the social agenda with its specific reference to supporting disadvantaged groups. However, to be able to take a stand on such policy priorities, it is clearly important for learning technologists to reflect upon what we value, both individually and collectively.

The invisibility of embedding

A dilemma facing learning technologists is that we can take no credit for our greatest successes. To be truly embedded, to become part of mainstream practice, technology needs to become so commonplace it is taken for granted. Effectively, it becomes invisible; it is like the tacit knowledge that is so essential to any form of professional practice (McMahon, 2000).

Once a new technology is familiar, it slips from focus, drifting into obscurity on the fringes of our field of vision – often at the same time some new technology occupies the foreground. Echoing Conole's chapter and looking back to the first issue of ALT-J, there are articles about the use of icons, others about hypertexts; typically, these are taken for granted in current design practice. Similarly, academics' uptake of email was once a topic for research. What university now would consider abandoning it? The situation is still more acute for non-computer-based technologies, such as overhead projectors, tape recorders (the subject of study in early volumes of the British Journal of Educational Technology), flipcharts or even the layout of lecture theatres.

It is striking, in reviewing these chapters, how the commonplace is conspicuously absent. Even in this review of the last decade, we find it hard to leave current struggles aside to see what has already been achieved. Thus I don't agree with Jacob's conclusion that in ten years' time, ALT's members will be "fine-tuning", "the war essentially having been won". The battle for online learning might be won, but another conflict will emerge, and ALT members are likely to focus their attention there whilst "fine tuning" online learning becomes part of someone else's routine duties. (Calverley, whose own chapter is more forward-looking than retrospectiv, identifies mobile computing as one such upcoming struggle.) Unless we strive to be more self-aware, unless we become more attentive to our own history, this pattern is set to continue.

An example of such risks can be found in Boyle & Cook's chapter on reuse (Chapter 2). Wisely, they challenge the received wisdom that "there is a direct inverse relationship between size and reusability", arguing instead for "learning microcontexts". Surprisingly, however, they do not consider the ubiquitous technologies that contradict the 'size rule', such as textbooks, classrooms or overhead projectors. Such objects clearly have a role to play in learning, are widely reused across sectors, nations and disciplines, yet fail to conform to recommendations about granularity (and in some cases have little direct link to 'content' at all). Whilst we focus on the visible problems of standards for content sharing, the cultural literacy problems facing potential users of such resources are ignored. Perhaps there is little difference between blanket advocacy of object economies and the management decrees criticised by Conole that all courses should use a Virtual Learning Environment (Chapter 10).

This fixation on the visible, and on its potential, is understandable but unhelpful. For example, we currently have quite a poor understanding of how teachers reuse these traditional, ubiquitous resources when designing curricula (Oliver, 2003). So is the object economy a problem experienced by teachers, or a vision expounded by advocates of a new technology? It may, of course, be both, but evidence and analysis is lacking. Similarly, what will the impact of this be on the teachers? If the creation of course reading lists is a reflection of a teacher's sense of individuality and of academic competence (Millen, 1997), will this be seen as liberation from routine tasks or 'dumbing down' a process that establishes them as experts? Is this really in the interests of learners, or is it what Oliver, O'Donoghue and Harper describe as being "more about flexible teaching than flexible learning"?

It is not just academics' practice we remain blind to: it is also our own. Littlejohn and Peacock describe a pioneering era a decade ago when enthusiastic IT staff provided technical support, then the evolution of this into pedagogically-focused partnerships. It is ironic that the difficulties of such a partnership approach are well described by Lawless & Kirkwood in a paper from 1976. Whether forms of practice became invisible or were simply forgotten is debatable, but there are clearly areas where we have failed to learn from our own history. Similarly, although the evolutionary model they present is generally convincing, it does not address the fact that several of these 'phases' now co-exist. For example, mainstream learning technology may no longer focus on technical support, but workshops on Dreamweaver and Access are widespread within many institutions, they are simply offered by IT trainers or Staff Development rather than by learning technologists.

Wilson's chapter on further education provides an interesting contrast. Even access to a PC, let alone to a networked machine, was problematic for teachers a decade ago, throwing the situation in higher education into sharp relief and reminding us how things taken for granted were still issues for colleges. Here, the struggle remains within memory, which helps keep technology in the foreground, providing a perspective on those situations lost to those of us who now take such resources for granted. And this balance is important, for as Conole discusses, if the technology is visible but the issues are not, all manner of inappropriate demands about its use may be made of practitioners.

Failure, maturity and theorising practice
The project-based, short-term, accountability-driven nature of learning technology funding represents a problem for researchers, as Dempster & Deepwell discuss. To gain another short-term contract, it is better to be associated with success than with failure. Jacob's observation that those involved with learning technology are biased towards promoting their projects, even when faced with patchy evidence, is thus entirely understandable. However, such individual short-term gains represent a problem for the field as a whole. They do little, for example, to help establish the recognition and kudos Conole asserts is so important.

Many assertions are made about e-learning, often in the form of policy statements. It may sound strange to label policy statements as 'theories', but this is what they are: statements of belief about how the world works. The same is true of the "factors influencing such success [that] are well known to educational developers across the sector, but have rarely been documented", in Dempster & Deepwell's discussion of project success.

That they are theories does not make them good, helpful or right, however. Indeed, our failures are an essential part of the process of understanding e-learning: they lie at the heart of theory-building. One way of moving beyond naïve theories is to follow the classic scientific process of thesis, antithesis and synthesis. The thesis might be that online, collaborative commercial education is a viable form of education; the experience of failed providers proves that it is not. But if we stop there, if we do not work out why the provider failed, so as to articulate our antithesis, we have no hope of synthesising these statements to produce new, more robust knowledge about learning technology. All that remains are the 'fashions' Jacobs alludes to, where each new acronym applied to the field allows it to be re-invented by policy-makers as a panacea for educational dilemmas, with no hope of deepening our understanding of its real effectiveness, or its limitations. It is striking in Wilson's chapter that individuals in further education seemed better able to admit to, and learn from, their failures than their colleagues in higher education.

This superficial reinvention raises a note of concern in relation to Calverley's discussion of project management. This is presented as a way of ensuring "effective development of courses and delivery of student obligations", a familiar call in this era of accountability. However, such patterns of work are about efficiency, not innovation. If we are obligated to succeed, how are we to justify the risks of fresh thinking? How are we to learn from failure (for some developments must surely fail) if such a thing becomes taboo?

The need for theory is particularly vivid in Calverley's own assertion that "the open commercial e-learning market is deemed to be of questionable value and quality by educators". Her contrast between the pedagogic model of such material, content presentation, with theoretically informed perspectives is telling. These theories are not abstract niceties; they are statements of belief about the way the world works. Generally, learning technologists just do not believe the 'default', transmissive model of education illustrated by these commercial offerings. They believe that learning arises from thoughtful experimentation (experiential learning), from questioning (critical thinking), from the intertwining of practice and reification, debated with peers (communities of practice). By deeming transmissive e-learning to be "of questionable value", we have taken a theoretical stand – but are we, individually and collectively, aware of what stand we have taken? Are we using these theories superficially to dismiss someone else's position, or have we made a commitment that this or that theory is, indeed, what we believe?

Ironically, whilst we espouse such theories, we are remarkably inconsistent when working with them. Behaviourism is rarely treated as credible, yet a remarkable amount of research can be explained purely in behaviourist terms. The skills training technologies Calverley describes, "where feedback precisely at the point of error allows more powerful and accurate association of that feedback with the thought process [...] and therefore greater ability to correct themselves", is essentially a drill-and-practice teaching machine, pedagogically identical to developments from the 1950s. Similarly, Boyle & Cook point out often much learning object design holds such "outdated" pedagogy. Yet if such perspectives are still proving powerful, perhaps it is time to abandon our fashions and begin to consider the values and shortcomings of each.

At present, learning technology draws on a proliferation of theories, many drawn from other disciplines and fields of study (Oliver & Aczel, 2002). This, I believe, would be a productive and generative position to maintain. Nonetheless, there is an important difference between individuals' commitment to a diversity of theories (offering an opportunity for productive debate) and general vagueness over which theories are deemed credible.

Teaching, learning, content and technology

There is an interesting contrast in perspectives within this book on the relationship between learning and resources. On the one hand, Boyle & Cook describe learning objects, a specific kind of resource, as "the basic reusable unit of learning"; on the other, Calverley argues that teaching is much more than simply providing resources, and learning more than experiencing teaching. To some extent, this is a matter of perspective: when designing materials it is convenient to ignore the problems associated with transforming intentions into experiences, because until these resources are actually used there is no hope of knowing what individuals will learn from them (Wenger, 1998). However, it is also important to remember that this position is an over-simplification, a conflation of very different relationships to the same object, a point Boyle & Cook make indirectly by highlighting three interpretations of 'learning objects' that are all held currently.

As discussed above, explaining and maintaining such distinctions requires a theoretical position. Yet even when we are clear these are separate, it is informative to consider how rhetoric emphasises one part or another. For example, Calverley's discussion of learning specifications and reference models highlights how discretely held information becomes part of an increasingly complex conception of 'content'; it also proposes that this should be "aligned across many new areas that belong to the process of reintegrating the institutional core business". This discussion may originate in concerns for student-centred course design, but its outcome is the subjugation of diverse but established institutional practices in order to facilitate content management. Whilst she debates whether "content is king", it is presented here as more important than the administrators and academics who traditionally worked with it. The same perspective is echoed by Boyle & Cook who, when considering barriers to reuse, conclude that it is the culture that must change, rather than (say) our expectations about learning objects.

Interestingly, Jacob's example of allLearn.org is described as an "international pooling of resources", yet a cursory glance reveals that resources are far from the heart of this initiative. Structured, facilitated discussions form the core of each course, supplemented by resources, typically traditional texts alongside videoed lectures. The format of resources may be new, but such pedagogy is far from innovative, a situation also observed by De Boer, Boezerooy and Fisser in the Netherlands (Chapter 7). Perhaps, however, there is good reason for this. Boyle & Cook discuss the problem of working with statisticians to develop a learning object that could teach a specific concept, finding that this group failed to avoid bringing in related concepts. An alternative to viewing this as a failure in clarity is to consider whether, for statisticians, the meaning of the concept can only be grasped when considered in relation to other concepts. This provides a significant shift in perspective: rather than focusing on isolated, packaged content to be internalised, it re-frames learning as the process of appreciating difference and of learning from the relationships, contrasts and juxtapositions of concepts. Rather than providing well-organised content, perhaps allLearn's discursive, negotiated approach (and the "stretching", rather than radical change, described by De Boer, Boezerooy and Fisser) works precisely because such (traditional) relational approaches to knowledge can be preserved. However, even this re-framing falls short of the important educational outcomes identified by Oliver, O'Donoghue and Harper when discussing what students learn from mixing with people from other races and cultures. Such serendipitous outcomes (such as mutual understanding and empathy) remain impossible to plan, let alone commodify.

It is also useful to note the division inherent in such discussions between content and the medium in which it is expressed. It is well established that there is a difference between data, information, knowledge and wisdom (e.g. Barnett, 1994). Moreover although (as Oliver, O'Donoghue and Harper point out) the influence of the medium on content has been widely debated, it is arguable that there is no 'message' distinct from the medium in which it is expressed. Taken together, we see evidence of an ongoing trend in contemporary society: what was personal knowledge is transformed into data for systems (Lyotard, 1979). The result is that what was once meaningful to people is transformed into artefacts, reified, in Wenger's terms (1998), in digital form.

There is a dilemma within this. To illustrate: Calverley presents two scenarios, one in which content is used to initiate and direct learning, the other in which it is treated as a resource to support self-directed learning. This echoes an ongoing worry within the field (described by Oliver, O'Donoghue and Harper) about whether learning technology is learning-led or technology-led. Such a dilemma is irresolvable, as long as we view these as two separate things. The alternative is to view learning technology as the *relationship between* learning and technology. (This perspective is inherent in Activity Theory, e.g. Kuutti, 1996, which is proving increasingly popular in learning technology research.) This perspective helps explain Calverley's call for "a better understanding of what each [form of delivery] is strong at." It also renders the term 'blended learning' meaningless, since there is no longer a clear-cut division between two separate 'kinds' of learning that need to be blended. The links between teaching, learning, content and technology are explicable in terms of a repertoire of teaching practices (each a relationship between a technology and what we intend students to learn) understood, and researched, in terms of their ability to foster relationships between cultural artefacts (such as specific pieces of media-specific 'content') and students. What still needs to be added to this picture, however, is what Boyle & Cook briefly allude to: the motivational, affective aspect of the process that leads students into such engagement.

Treating teachers as learners

As discussed above, the fashion amongst learning technologists is to favour constructivist, experiential theories of learning over transmissive approaches. Sometimes, however, it seems we lose this perspective when working with our colleagues. I argued earlier that learning technology can be interpreted as a repertoire of practice, fostering relationships between students and content. If this model is credible, it becomes unsurprising that Calverley identifies a retrograde step in teaching practice when new technologies, such as the Web, are introduced. Stepping back from interactive multimedia to passive text-based pages is unfortunate, but should be interpreted in terms of teachers making sense of their unfamiliar relationships with new types of technology. They cannot 'learn' simply by being told what this technology is, they have to gain experience and contextualise it within their own practice in order to make sense of it. And, of course, doing so changes their practice, expanding their repertoire and thus supporting new ways of fostering students' engagement with content. This helps explain why De Boer, Boezerooy and Fisser find that redesign starts with the course as instructors already know it and that teachers then begin "stretching the mold" – for where else can they begin, and how else should they proceed? Similarly, it offers a new perspective on the "false assumption" identified by Littlejohn and Peacock that "exposure to computers and CAL packages was sufficient to drive the development of new forms of teaching with technology." Rather than being a sufficient condition (i.e. this would *cause* new forms of teaching), it could be re-interpreted as a *necessary* condition (i.e. it doesn't cause new forms by itself, but new forms won't arise without this happening at some point). Initially, some of the new approaches will be unsuccessful: as argued earlier, these failures form a necessary part of this learning (and theorising) process. These are the kind of activities that, in Wilson's terms, are about "pushing the boundaries" rather than simply replicating existing practice.

Of course, this runs counter to the dominant 'efficiency' model, "the current trend" identified by Calverley, involving removing development from academics. Under such an approach, teachers are not learners, they are simply content experts, kept at arm's length from the specialist process of instructional design (Merrill, 2001). Perhaps this approach is more efficient, but is it as useful? If the medium influences students' engagement, and it is clear that tasks rarely work the same way online as face-to-face, how should teachers develop their repertoire of practices? How can they make sense of a new technology they are excluded from? How, can they move from "e-replication" to more sophisticated, innovative uses of technologies? How are they to avoid the situation described by Wilson, where everyone's course looks the same, a position that "can be hugely gratifying for college quality managers and make the processing of returns to funding bodies easier, but it might not be that rewarding for students or tutors in the long run"?

Dempster & Deepwell's chapter evidences the success of involvement, where providing a "supportive playground" enabled teachers to gain experience, gain knowledge and then make decisions or commit resources to personal and institutional initiatives. This is vital. As Dempster & Deepwell identify, initiatives do not simply appear: they are put in place by people. Where individuals develop their understanding of this field and then move into committees, they promote their perspective through the policies and activities they help design, influencing future change in ways connected to practice and a real understanding of what learning technology means.

This sense of individual progression provides an alternative interpretation to the historical perspective Littlejohn & Peacock describe. Their model moves from pioneering, through diversification, to formalisation, then a focus on pedagogical effect and finally partnership. This seems well suited to describing individuals' engagement with the process of learning through experience. This seems so credible as an historical perspective simply because there is a sizable group progressing through this process at a comparable rate, arising from posts funded under TLTP and other such initiatives. In effect, this group has become a 'cohort' that has progressed through this 'curriculum' together. However, it seems likely that individuals new to this field, for example, academics that have only recently begun to engage with learning technology, will follow comparable steps themselves. This model thus not only describes our shared experience to date, but could also act as a model for future academic development.

It is also interesting to note the similarities between this and Conole's description of foci for research, with different areas considering the creation of new technologies, their use, their impact on practice, their wider impact on the organisation and so on. Again, the fact that these theory-building, meaning-making phases occur in parallel, not sequentially, suggests that people may need to move 'up' and 'down' Littlejohn & Peacock's model over time as new technologies emerge. They may, to draw on Dempster & Deepwell's analogy, need a new "supportive playground" for each emerging technology.

Again, then, we are faced with a dilemma for which there is no simple 'solution'. Instead of deciding whether teachers should or should not be developers, this becomes a matter of degree: how much involvement do teachers need in the developmental process to ensure that an appropriate balance between their learning and institutional efficiency is struck?

Conclusions

The past decade has witnessed considerable progress within the field of learning technology, yet it is clear that we remain only dimly self-aware, often preoccupied with current difficulties. This situation, which is exacerbated by the short-term contracts many of us are on, leaves little room for reflection, enhancing a tendency for us to repeat the forgotten patterns of the past.

Yet there is evidence of a growing self-awareness. We recognise the complexity of organisational change and of learning, whether it involves students, peers or ourselves. We are developing theory, enabling us to understanding what we do. We have started to challenge the policies we believe are misguided or inappropriate. But more needs to be done. This is where my personal agenda becomes apparent. Others may disagree about priorities, but based on my reflection on these chapters and my reflexion upon my own role in this work, there are several areas that I see as priorities for the coming decade.

Rather than presenting a simple binary between 'e-learning' and 'learning', we must recognise that all learning involves technology, it is simply a matter of how familiar we are with it. If we are too familiar, it will slip from view. If we are to help others work with new technologies, we must avoid projecting our own expectations onto them. This involves finding ways of making taken-for-granted technologies visible, so that we can re-interpret these practices not as simple skills ("have we taught them this or not?") but as forms of cultural literacy. Littlejohn and Peacock argue that technology has become ubiquitous in everyday life: certain technologies have, but others have not. To do justice to our field, we need to balance our fascination with the new and strange (such as mobile computing) alongside the overly familiar.

We also need to develop our own understanding. Learning technology involves working across numerous communities, disciplinary, technical, managerial and so on. Each will understand technology differently. This is what makes learning technology exciting: contrasts and conflicts between different theories and beliefs has proved productive and generative. If we are to continue our meaning-making work, helping people to understand what learning technology means, we must continue to struggle to connect communities and value the diverse perspectives they will bring. We must also, however, come to terms with our own values. Conole rightly identifies the growing political awareness within research and practice. To respond to this we must take a stand on what we believe is important, as individuals and as a community. Rather than simply provide a service, implementing policies that are handed to us, we should develop ourselves as a *critical* service: questioning and challenging initiatives we believe to be ill-informed or inappropriate. We should learn to form judgements about the value of the work we do, judging whether it is 'good practice' not simply in terms of efficiency (economics) but also on the basis of its educational, social or moral impact. If learning technology is to be recognised as a professional activity, its practitioners must be able to consider their work from an ethical *as well as* an economic perspective.

References

Barnett, R. (1994). *Limits of Competence: Knowledge, Higher Education and Society*. Buckingham: Open University/SRHE press.

Beetham, H., Jones, S. and Gornall, L. (2001), *Career Development of Learning Technology Staff: Scoping Study Final Report*, JISC Committee for Awareness, Liaison and Training Programme. [On-Line]. Available: *http://sh.plym.ac.uk/eds/effects/jcalt-project/*.

Boyer, E. (1990). *Scholarship Reconsidered: Priorities of the Professoriate*. London: Josey-Bass.

Conole, G. (2003). The evolving landscape of learning technology. *ALT-J*, 10, 3, 4-18.

DFES (2003), *The Future of Higher Education*, London: Department for Education and Skills. *[On-Line]*. Available: www.dfes.gov.uk/highereducation/hestrategy

Holley, D.,& Oliver, M. (2000). Pedagogy and new power relationships. *International Journal of Management Education*, 1,1, 11-21.

Jones, C. (2002*). Is There a Policy for Networked Learning*? In Proceedings of the Networked Learning Conference, Lancaster. http://www.shef.ac.uk/nlc2002/proceedings/symp/08.htm#08a.

Kuutti, K. (1996) Activity theory as a potential framework for human-computer interaction research. In Nardi, B. (Ed.) *Context and consciousness: activity theory and human-computer interaction*, 17-44. Cambridge: MIT Press.

Lawless, C., & Kirkwood, A. (1976). Training the Educational Technologist. *British Journal of Educational Technology*,1,7, 54-60.

Lisewski, B. & Joyce, P. (2003) Examining the five-stage e-moderating model: designed and emergent practice in the learning technology profession. *ALT-J*, 11, 1, 55-66.

Lyotard, J-F. (1979). *The Postmodern Condition: A Report on Knowledge.* Manchester: Manchester University Press.

Marsick, V., & Watkins, K. (1999). Envisioning new organisations for learning. In D. Boud, & Garrick, J. (Eds.), *Understanding Learning at Work* (pp199-215). London: Routledge.

McMahon, A. (2000). The development of professional intuition. In T. Atkinson & G. Claxton (Eds), *The Intuitive Practitioner: On the Value of Not Always Knowing What One is Doing* (pp 137-148). Buckingham: Open University Press.

Merrill, M. (2001). Components of instruction: toward a theoretical tool for instructional design. *Instructional Science*, 29,4, 291-310.

Millen, J. (1997). Par for the Course: designing course outlines and feminist freedoms. *Curriculum Studies*, 5,1, 9-27.

NCIHE - National Committee of Inquiry into Higher Education (1997) *Higher Education in the Learning Society.* London: HMSO.

Oliver, M., & Aczel, A. (2002). A commentary on the use of theory in the analysis of the Jape study. Appendix to, Theoretical Models of the Role of Visualisation in Learning Formal Reasoning. *Journal of Interactive Media in Education.* [On-Line]. Available: http://www-jime.open.ac.uk/2002/3/

Oliver, M. (2003). *Curriculum Design as acquired social practice: a case study of academics in UK Higher Education.* Paper presented at the 84th Annual Meeting of the American Educational Research Association, Chicago. [On-Line]. Available:
http://www.tigersystem.net/aera2003/revviewproposaltext.asp?propid=3352

Oliver, M. (2002) What do learning technologists do? *Innovations in Education and Teaching International*, 39,4, 1-8.

Smith, H. & Oliver, M. (2002).University teachers' attitudes to the impact of innovations in ICT on their practice. In Rust, C. (Ed.), *Proceedings of the 9th International Improving Student Learning Symposium* (pp 237-246). Oxford: Oxford Centre for Staff and Learning Development.

Taylor, P. (1999). *Making sense of academic life: Academics, Universities and Change.* Buckingham: Open University/SRHE press.

Thorpe, M. (2002) From independent learning to collaborative learning: new communities of practice in open, distance and distributed learning. In M. Lea & K. Nicoll (Eds), *Distributed Learning: Social and Cultural Approaches to Practice*, 131-151. London: RoutledgeFalmer.

Wenger, E. (1998) *Communities of Practice.* Cambridge: Cambridge University Press.

Contributors

Petra Boezerooy
Centre for Higher Education Policy Studies (CHEPS)
University of Twente
7500 AE Enschede
Netherlands

Tom Boyle
Learning Technology Research Institute
London Metropolitan University
London E2 8AA
United Kingdom

Gayle Calverley
Distributed Learning
University of Manchester
Manchester M13 9GP
United Kingdom

Gràinne Conole
School of Education
University of Southampton
Southampton SO17 1BJ
United Kingdom

John Cook
Learning Technology Research Institute
London Metropolitan University
London E2 8AA
United Kingdom

Wim de Boer
Faculty of Educational Science and Technology
University of Twente
7500 AE Enschede
The Netherlands

Frances Deepwell
Centre for Higher Education Development
Coventry University
Coventry CV1 5FB
United Kingdom

Jacqueline Dempster
Centre for Academic Practice
University of Warwick
Coventry CV4 7AL
United Kingdom

Petra Fisser
Digitale Universiteit
3500 AD Utrecht
The Netherlands

Barry Harper
University of Wollongong
Wollongong
NSW 2522
Australia

Gabriel Jacobs
European Business Management School
University of Wales Swansea
Swansea SA2 8PP
United Kingdom
Allison Littlejohn
Centre for Academic Practice
University of Strathclyde
Glasgow G1 1QE
United Kingdom

John O'Donoghue
Centre for Learning & Teaching
University of Wolverhampton
Wolverhampton WV1 1SB
United Kingdom

Martin Oliver
Education and Professional Development
University College London
London WC1E 6BT
United Kingdom

Ron Oliver
School of communications and Multimedia
Edith Cowan University
Mt Lawley 6050
Western Australia

Susi Peacock
Centre for Academic Practice
Queen Margaret University College
Edinburgh EH12 8TS
United Kingdom

Jane Seale
School of Education
University of Southampton
Southampton SO17 1BJ
United Kingdom

Joe Wilson
Scottish Further Education Unit
Stirling FK9 4TY
United Kingdom

Index